Applied Analytics through Case Studies Using SAS and R

Implementing Predictive Models and Machine Learning Techniques

Deepti Gupta

Apress®

Applied Analytics through Case Studies Using SAS and R

Deepti Gupta
Boston, Massachusetts, USA

ISBN-13 (pbk): 978-1-4842-3524-9   ISBN-13 (electronic): 978-1-4842-3525-6
https://doi.org/10.1007/978-1-4842-3525-6

Library of Congress Control Number: 2018952360

Managing Director, Apress Media LLC: Welmoed Spahr
Acquisitions Editor: Celestin John
Development Editor: James Markham
Coordinating Editor: Divya Modi

Cover designed by eStudioCalamar

Cover image designed by Freepik (www.freepik.com)

Distributed to the book trade worldwide by Springer Science+Business Media New York, 233 Spring Street, 6th Floor, New York, NY 10013. Phone 1-800-SPRINGER, fax (201) 348-4505, e-mail orders-ny@springer-sbm.com, or visit www.springeronline.com. Apress Media, LLC is a California LLC and the sole member (owner) is Springer Science + Business Media Finance Inc (SSBM Finance Inc). SSBM Finance Inc is a **Delaware** corporation.

For information on translations, please e-mail rights@apress.com, or visit http://www.apress.com/rights-permissions.

Apress titles may be purchased in bulk for academic, corporate, or promotional use. eBook versions and licenses are also available for most titles. For more information, reference our Print and eBook Bulk Sales web page at http://www.apress.com/bulk-sales.

Any source code or other supplementary material referenced by the author in this book is available to readers on GitHub via the book's product page, located at www.apress.com/978-1-4842-3524-9. For more detailed information, please visit http://www.apress.com/source-code.

Printed on acid-free paper

I am dedicating this book to my Family.

Table of Contents

TABLE OF CONTENTS

About the Author

Deepti Gupta completed her MBA in Finance & PGPM in Operation Research in 2010. She has worked with KPMG and IBM private limited as a Data Scientist and is currently working as a data science freelancer. Deepti has extensive experience in predictive modeling and machine learning and her expertise is in SAS and R. Deepti has developed data science courses and delivered data science trainings and conducted workshops in both corporate and academic institutions. She has written multiple blogs and white papers. Deepti has a passion for mentoring budding data scientists.

About the Contributor

 **Dr. Akshat Gupta** is currently working as a Senior Applications Engineer at MilliporeSigma in Applications Engineering, Global Manufacturing Sciences and Technology (MSAT) group. He authored the healthcare case study (Chapter5) of this book. His focal area of research is cell culture clarification and tangential flow filtration. Dr. Gupta has extensive experience in Design of Experiments (DOE) and statistical analysis. He holds a Bachelor of Technology (B.Tech) degree in Chemical Engineering from the Vellore Institute of Technology, and a Master of Science (MS) and Doctor of Philosophy (Ph.D.) in Chemical Engineering from the University of Massachusetts Lowell. He also has graduate certificates in Modeling and Simulation, and Nanotechnology.

About the Technical Reviewer

Preeti Pandhu has a Master of Science degree in Applied (Industrial) Statistics from the University of Pune. She is SAS certified as a base and advanced programmer for SAS 9 as well as a predictive modeler using SAS Enterprise Miner 7. Preeti has more than 18 years of experience in analytics and training.

She started her career as a lecturer in statistics and began her journey into the corporate world with IDeaS (now a SAS company), where she managed a team of business analysts in the optimization and forecasting domain. She joined SAS as a corporate trainer before stepping back into the analytics domain to contribute to a solution-testing team and research/consulting team. She was with SAS for 9 years. Preeti is currently passionately building her analytics training firm, DataScienceLab (`www.datasciencelab.in`).

Acknowledgments

Book writing is one of the most interesting and challenging attempt one can take up. This book could not have been completed without the encouragement, guidance, and support of my family. I would like to thank Dr. Akshat Gupta, Ved Prakash Garg, Col. Atul Gupta, Dr. Anvita Garg, Ayush Gupta, RS Miyan, Ansi Miyan, Dr. James Chrostowski, and my colleagues and friends for their productive discussions and suggestions. My special thanks to Celestin John who provided great help on everything ranging from technical support to answering my queries. I appreciate the thoughtful and insightful comments from the editor and the reviewers. Thanks to the Apress team, especially to Divya Modi for all the patience, support, and guidance in completing this project.

Introduction

Analytics is a big buzz and a need for today's industries to solve their business problems. Analytics helps in mining the structured and unstructured data in order to withdraw the effective insights from the data, which will help to make effective business decisions. SAS and R are highly used tools in analytics across the globe by all industries for data mining and building machine learning and predictive models. This book focuses on industrial business problems and a practical analytical approach to solve those problems by implementing predictive models and machine learning techniques using SAS and R analytical languages.

The primary objective of this book is to help statisticians, developers, engineers, and data analysts who are well versed in writing codes; have a basic understanding of data and statistics; and are planning to transition to a data scientist profile. The most challenging part is practical and hands-on knowledge of building predictive models and machine learning algorithms and deploying them in industries to address industrial business problems. This book will benefit the reader in solving the business problems in various industrial domains by sharpening their analytical skills in getting practical exposure to various predictive model and machine learning algorithms in six industrial domains.

What's in This Book

This book focuses on industrial business problems and practical analytical approaches to solve those problems by implementing predictive models and machine learning techniques using SAS Studio and R analytical languages. **This book contains six industrial case studies of various domains with data and all the codes in SAS Studio and R languages, which would benefit all readers to practice and implement these models in their own business cases.**

In Chapter 1 the general outline about analytics, the role of analytics in various industries, and a few popular data science and analytical tools are discussed. Chapter 2 describes the role of analytics in the banking industry with a detailed explanation of predicting a bank loan default case study in R and SAS. Chapter 3 describes how analytics contribute in the retail industry and offers a detailed explanation of forecasting a case study in R and SAS. Chapter 4 describes how analytics is reshaping the telecommunications industry and gives a detailed explanation of a case study on predicting customer churn in R and SAS. Chapter 5 describes the application of analytics in the healthcare industry and gives a clear explanation of a case study on predicting the probability of benign and malignant breast cancer using R and SAS. Chapter 6 describes the role of analytics in the airline industry and provides a case study on predicting flight arrival delays (minutes) in R and SAS. Chapter 7 describes the application of analytics in the FMCG industry with a detailed explanation of a business case study on customer segmentation based on their purchasing history using R and SAS.

Who's the Target Audience?

- Data Scientists who would like to implement machine learning techniques with a practical analytical approach toward a particular industrial problem.

- Statistician, Engineers, and Researchers with a great theoretical understanding of data and statistics and would like to enhance their skills by getting practical exposure to data modeling.

- Data analysts who know about data mining but would like to implement predictive models and machine learning techniques.

- Developers who are well versed with coding but would like to transition to a career in data science.

What You Will Learn

- Introduction to analytics and data understanding.

- How to approach industrial business problems with an analytical approach.

- Practical and hands-on knowledge in building predictive model and machine learning techniques.

- Building the analytical strategies.

CHAPTER 1

Data Analytics and Its Application in Various Industries

Data analytics has become part and parcel of any business in today's world. In fact, it has evolved into an industry in itself. Vast numbers of software platforms are available for data extraction, scrubbing, analysis, and visualization. Some of these platforms are specialized for carrying out one of the above-listed aspects of data analytics, while others offer a generalist tool to carry out almost all tasks ranging from data scrubbing to visualization. Of these platforms, SAS® and R are the most popular for data analytics with a large global clientele.

In 1967, Statistical Analysis System (SAS) started as a federal funded project for graduate students to track agriculture data at North Carolina State University.[1] Today it has become a global leader in data analysis software market with customers spanning over 148 countries.[2] Ninety-six of the top 100 Fortune Global 500® companies use SAS. R, which originally was a statistical computing language, has advanced significantly over the years. R Studio is an Integrated Development Environment (IDE) for R[3] and offers a free, user-friendly platform for data analytics. Both SAS® and R offer vast capabilities but have certain contrasting advantages that are discussed later in more detail.

A broad array of companies ranging from the largest global banks to regional transport firms are using data analytics to solve diverse sets of problems These diverse applications have one commonality: using data and statistics as the basis for decision making.

In this chapter, certain key aspects related to data analytics will be introduced.

© Deepti Gupta 2018
D. Gupta, *Applied Analytics through Case Studies Using SAS and R*,
https://doi.org/10.1007/978-1-4842-3525-6_1

What Is Data Analytics?

Analytics is defined as the process of developing the actionable insights through the application of statistical model and analysis from the data.[4] Applying data analytics for decision making is a systematic process. It starts with understanding the nature of industry, general functionality, bottlenecks, and challenges specific to the industry. It is also helpful to know who the key companies are, size of industry, and in some cases general vocabulary and terms associated with operations. After that we take a deeper dive in to the area specific to the application or a business case to which data analytics needs to be applied. A thorough understanding of the application, associated variables, sources of data, and knowledge of the reliability of different data sources are very important.

Data analytics firms pay a lot of attention to these aspects and often employ a vast number of subject-matter experts specific to industries and at times even specific to certain key applications. Business research consultants are also employed for gaining understanding and insights in certain cases. During the preliminary phase of a project, data analytics firms perform elaborate surveys and conduct series of interviews to gain more information about the company and the business problem.[5] A good understanding of industry and the application can result in significant cost saving and can improve accuracy, performance, and practicality of the model.

Once the application or the problem statement is well understood, then the implementation process starts. The core methodology of implementing data analytics for solving a business problem is shown in Figure 1-1.[6]

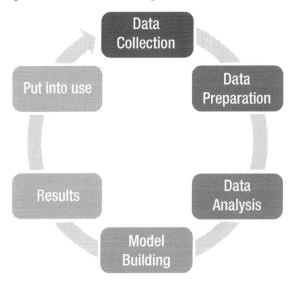

Figure 1-1. *Data Analytics Methodology*

Data Collection

The first step in the process is data collection. Data relevant to the applicant is collected. The quality, quantity, validity, and nature of data directly impact the analytical outcome. A thorough understanding of the data on hand is extremely critical.

It is also useful to have an idea about some other variables that may not directly be sourced from the industry or the specific application itself but may have a significant impact if included into the model. For example, when developing a model to predict flight delays, weather can be a very important variable, but it might have to be obtained from a different source then the rest of the data set. Data analytics firms also have ready access to certain key global databases including weather, financial indices, etc. In recent years, data mining of digital social media like Twitter and Facebook is also becoming very popular.[7] This is particularly helpful in understanding trends related to customer satisfaction with various services and products. This technique also helps reduce the reliance on surveys and feedbacks. Figure 1-2 shows a Venn diagram of various sources of data that can be tapped into for a given application.

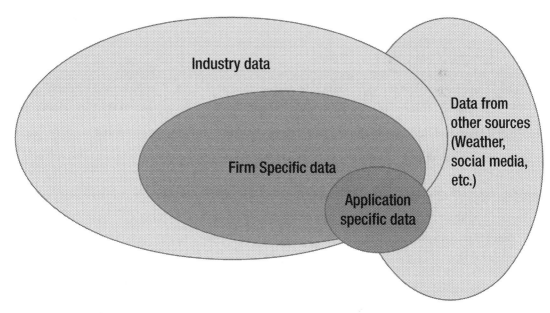

Figure 1-2. *Venn diagram of data sources*

Data Preparation

The next step is data preparation. Usually raw data is not in a format that can be directly used to perform data analysis. In very simple terms, most platforms require data to be in a matrix form with the variables being in different columns and rows representing various observations. Figure 1-3 shows an example of structured data.

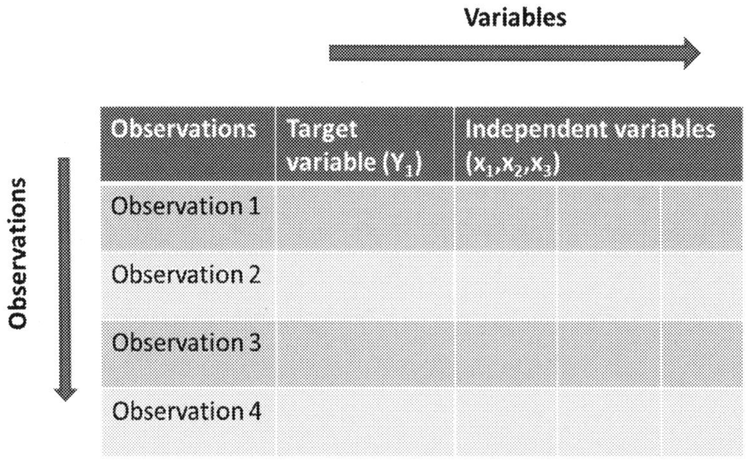

Figure 1-3. *Format of structured data*

Data may be available in structured, semi-structured, and unstructured form. A significant effort is needed to align semi-structured and unstructured data into a usable form as shown in Figure 1-3. Once the data is brought together in a structured form, the next stage in data preparation is data cleansing or scrubbing. Data scrubbing encompass processes that help remove inconsistencies, errors, missing values, or any other issues that can pose challenges during data analysis or model building with a given data set.[8] Work at this stage can be as simple as changing the format of a variable, to running advanced algorithms to estimate suitable estimates for missing values. This task is significantly more involved when it comes to big data.

Data Analysis

Once data is converted into a structured format, the next stage is to perform data analysis. At this stage underlying trends in the data are identified. This step can include fitting a linear or nonlinear regression model, performing principal component analysis or cluster analysis, identifying if data is normally distributed or not. The goal is to identify what kind of information can be extracted from the data and if there are underlying trends that can be useful for a given application. This phase is also very useful for scoping out the models that can be most useful to capture the trends in data and if the data satisfies underlying assumptions for the model. One example would be to see if the data is normally distributed or not to identify if parametric models can be used or a non-parametric model is required.

Model Building

Once the trends in data are identified, the next step is to put the data to work and build a model that will help with the given application or help solve a business problem. A vast number of statistical models are available that can be used, and new models are being developed every day. Models can significantly vary in terms of complexity and can range from simple univariate linear regression models to complex machine learning algorithms. Quality of a model is not governed by complexity but rather by its ability to account for real trends and variations in data and sift information from noise.

Results

Results obtained from the models are validated to ensure accuracy and model robustness. This can be done two ways; the first is by splitting the original data set into training and validation data sets. In this approach, part of the data is used for model building and the remaining part is used for validation. The other approach is to validate data against real-time data once the model is deployed. In some cases, the same data is used to build multiple different types of models to confirm if the model outputs are real and not statistical artifacts.

Put into Use

Once the model is developed it is deployed in a real-time setting for a given application. As shown in the Figure 1-1, the overall process is somewhat iterative in nature. Many times, the models have to be corrected and new variables added or some variables removed to enhance model performance. Additionally, models need to be constantly recalibrated with fresh data to keep them current and functional.

Types of Analytics

Analytics can be broadly classified under three categories: descriptive analytics, predictive analytics, and prescriptive analytics.[9] Figure 1-4 shows the types and descriptions of types of analytics.

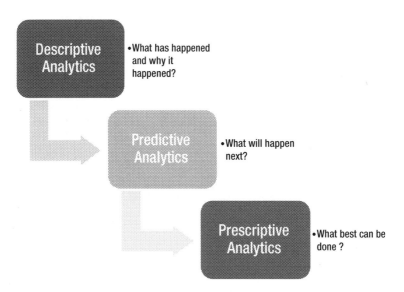

Figure 1-4. *Types of Analytics*

Different types of information can be obtained by applying the different categories of analytics. This will be explained in the following section.

1. **Descriptive Analytics**: Most of the organizations use descriptive analytics in order to know about their company performance. Example, management at a retail firm can use descriptive analytics to know the trends of sales in past years, or inferring trends of operation cost, product, or service performance.

2. **Predictive Analytics**: In case of predictive analytics, historical trends coupled with other variables are used to see what could happen in the future to the firm. Example, Management at the same retail firm can use the sales trends from previous years to forecast sales for the coming year.

3. **Prescriptive Analytics**: In prescriptive analytics, the objective is to identify factors or variables that are impacting trends. Once the responsible variables are identified, strategies and recommendations are made to improve the outcome. For example, Management at the same retail firm identifies that the operation cost is significantly high due to overstocking at certain stores. Based on this insight, an improved inventory management would be recommended to the given locations.

Understanding Data and Its Types

Data is a collection of variables, facts, and figures that serves as raw material to create information and generate insights. The data needs to be manipulated, processed, and aligned in order to withdraw useful insights. Data is divided into two broad forms: qualitative and quantitative data.[10]

1. **Qualitative data**: The data that is expressed in words and descriptions like text, images, etc. is considered as qualitative data. Qualitative data collection uses unstructured and semi-structured techniques. There are various common methods to collect qualitative data like conducting interviews, diary studies, open-ended questionnaires, etc. Examples of qualitative data are gender, demographic details, colors, etc. There are three main types of qualitative data:

   - **Nominal**: Nominal data can have two or more categories but there is no intrinsic rank or order to the categories. For example, gender and marital status (single, married) are categorical variables having two categories and there is no intrinsic rank or order to the categories.

   - **Ordinal**: In ordinal data, the items are assigned to categories and there is an intrinsic rank or order to the categories. For example, age group: Infant, Young, Adult, and Senior Citizen.

   - **Binary**: Binary data can take only two possible values. For example, Yes/No, True/False.

2. **Quantitative data**: The data that is in numerical format is
   considered as quantitative data. Such a type of data is used in
   conducting quantitative analysis. Quantitative data collection
   uses much more structured techniques. There are various
   common methods to collect quantitative data like surveys,
   online polls, telephone interviews, etc. Examples of quantitative
   data are height, weight, temperature, etc. There are two types of
   quantitative data:

   - **Discrete Data**: Discrete data is based on count and it can only
     take a finite number of values. Typically it involves integers.
     For example, the number of students in data science class is
     discrete data because you are counting a whole and it cannot be
     subdivided. It is not possible to have 8.3 students.

   - **Continuous Data**: Continuous data can be measured, take any
     numeric values, and be subdivided meaningfully into finer and
     finer levels. For example, the weights of the data science students
     can be measured at a more precise scale – kilograms, grams,
     milligrams, etc.

While on the topic of data, it is a good time to get a basic understanding of "Big Data."
Big Data is not just a buzzword but is fast becoming a critical aspect of data analytics.
It is discussed in more detail in the following section.

What Is Big Data Analytics?

The term "big data" is defined as the huge volume of both structured and unstructured
data that is so large that it is not possible to process such data using traditional databases
and software. As a result, many organizations that collect, process, and conduct big
data analysis turn to specialized big data tools like NoSQL databases, Hadoop, Kafka,
Mapreduce, Spark, etc. Big data is a huge cluster of numbers and words. Big data
analytics is the process of finding the hidden patterns, trends, correlations, and other
effective insights from those large stores of data. Big data analytics helps organizations

harness their data to use it for finding new opportunities, faster and better decision making, increased security, and competitive advantages over rivals, such as higher profits and better customer service. Characteristics of Big data are often described using 5 Vs, which are velocity, volume, value, variety, and veracity.[11] Figure 1-5 illustrates 5 Vs related to the big data.

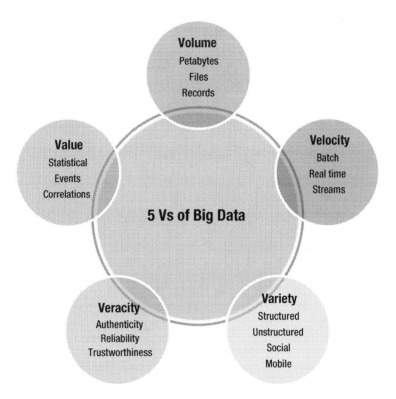

Figure 1-5. *5 Vs of Big Data*

Big Data analytics applications assist data miners, data scientists, statistical modelers, and other professionals to analyze the growing volumes of structured and mostly unstructured data such as data from social media, emails, web servers, sensors, etc. Big data analytics helps companies to get accessibility to nontraditional variables or sources of information, which helps organizations to make quicker and smarter business decisions.

Big Data Analytics Challenges

Most of the organizations are experiencing effective benefits by using big data analytics, but there are some different obstacles that is making it difficult to achieve the benefits promised by big data analytics.[12] Some of the key challenges are listed below:

- Lack of internal skills: The most important challenge that organizations face in implementing big data initiatives is lack of internal skills, and there is a high cost of hiring data scientists and data miners for filling the gaps.

- Increasing growth of the data: Another important challenge of big data analytics is the growth of the data at a tremendous pace. It creates issues in managing the quality, security, and governance of the data.

- Unstructured Data: As most of the organizations are trying to leverage new and emerging data sources, it is leading to the more unstructured and semi-structured data. These new unstructured and semi-structured data sources are largely streaming data coming from social medial platforms like Twitter, Facebook, web server logs, Internet of Things (IOT), mobile applications, surveys, and many more. The data can be in the form of images, email messages, audio and video files, etc. Such unstructured data is not easy to analyze without having advanced big data analytical tools.

- Data Siloes: In organizations there are several types of applications for creating the data like customer relationship management (CRM), supply chain management (SCM), enterprise resource planning (ERP), and many more. Integrating the data from all these wide sources is not an easy task for the organization and is one of the biggest challenges faced by big data analytics.

Data Analytics and Big Data Tools

Data science and analytics tools are evolving and can be broadly classified into two classes: tools for those techies with high levels of expertise in programming and profound knowledge of statistics and computer science like R, SAS, SPSS, etc.; and tools for common audiences that can automate the general analysis and daily reports like Rapid Miner, DataRPM, Weka, etc. Figure 1-6 displays the currently prevalent languages, tools, and software that are used for various data analytics applications.

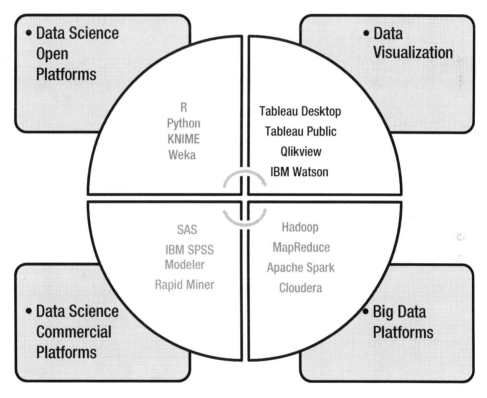

Figure 1-6. *Languages, Tools, and Software*

There is a long list of tools, and a few popular data science and analytical tools are discussed in the following section.

1. **R: The Most Popular Programming Language for statisticians and data scientists**

   R is an open source tool widely used by statisticians and data miners for conducting statistical analysis and modeling.[13] R has thousands of packages available easily that make the jobs of statisticians and data scientists easy for handling the tasks from text analytics to voice recognition, face recognition, and genomic science. The demand of R has increased dramatically across all the industries and is becoming popular because of its strong package ecosystem. R is used in industries for solving their big data issues and building statistical and predictive models for withdrawing the effective insights and hidden patterns from the data.

2. **SAS (Statistical Analysis System) Data Science and Predictive Analytics Software Suite**

   SAS is a software suite that is popular for handling large and unstructured data sets and is used in advance analytics, multivariate analysis, data mining, and predictive analytics, etc. The SAS software suite has more than 200 components like BASE SAS, SAS/ STAT, SAS/ETS, SAS/GRAPH, etc. BASE SAS software, SAS Enterprise Guide, and SAS Enterprise Miner are licensed tools and are used for commercial purposes by all the industries. SAS University Edition is free SAS software and is used for noncommercial uses like teaching and learning statistics and modeling in an SAS environment. It includes the SAS components BASE SAS, SAS/STAT, SAS/IML, SAS/ACCESS, and SAS Studio. SAS can be expensive but it is a very popular tool in industries; it has an effective and quick support system and more than 65,000 customers.

3. **IBM SPSS Statistics and SPSS Modeler: Data Mining and Text Analytics Software**

    SPSS Modeler and SPSS Statistics were acquired by IBM in 2009 and is considered as a data mining, statistical, and text analytics software. It is used to load, clean, prepare the data, and then build the predictive models and conduct other analytical and statistical tasks. It has the visual interface so users without good programming knowledge can easily build the predictive model and statistical analysis.[14] It has been widely used in industries for fraud detection, risk management, forecasting, etc. IBM SPSS modeler (version 17) is present in two separate bundles as:

    1. SPSS Modeler Professional: it is used for structured data such as databases, flat files, etc.

    2. SPSS Modeler Premium: it is a high-performance analytical tool that helps in gaining effective insights from the data. It includes all the features from SPSS Modeler Professional and in addition it is used for conducting Text Analytics,[15] Entity Analytics,[16] and Social Network Analytics.

4. **Python: High-Level Programming Language Software**

    Python is an object-oriented and high-level programming language.[17] Python is easy to learn and its syntax is designed to be readable and straightforward. Python is used for data science and machine learning. Robust libraries used for data science and machine learning are using the interface of Python, which is making the language more popular for data analytics and machine learning algorithms.[18] For example, there robust libraries for statistical modeling (Scipy and Numpy), data mining (Orange and Pattern), and supervised and unsupervised machine learning (Scikit-learn).[19]

5. **Rapid Miner: GUI Driven Data Science Software**

   Rapid Miner is open source data mining software. It was started in 2006 and was originally called Rapid-I. In 2013 the name was changed from Rapid-I to Rapid Miner. The older version of Rapid Miner is open source but the latest version is licensed. Rapid miner is widely used in industries for data preparation in visualization, predictive modeling, model evaluation, and deployment.[20] Rapid Miner has a user-friendly graphic user interface and a block diagram approach. Predefined blocks act as a plug and play system. Connecting the blocks accurately helps in building a wide variety of machine learning algorithms and statistical models without writing a single line of code. R and Python can also be used to program Rapid Miner.

Role of Analytics in Various Industries

The onset of the digital era has made vast amounts of data accessible, analyzable, and usable. This, coupled with a highly competitive landscape, is driving industries to adopt data analytics. Industries ranging from banking and telecommunication to health care and education, everyone is applying various predictive analytics algorithms in order to gain critical information from data and generate effective insights that drive business decisions.

There are vast numbers of applications within each industry where data analytics can be applied. Some applications are common across many industries. These include customer-centric applications like analyzing factors impacting customer churn, engagement, and customer satisfaction. Another big data analytics application is for predicting financial outcomes. These include forecasting of sales, revenues, operation costs, and profits. In addition to these, data analytics is also widely used for risk management and fraud detection and price optimization in various industries.

There are also large numbers of industry-specific applications of data analytics. To list a few: flight delay prediction in the aviation industry, prediction of cancer remission in health care, forecasting wheat production in agriculture.

An overview of some of the industries benefiting from predictive and big data analytics insights and, most importantly, how is discussed in this section.

1. **Insurance Industry:**

   The insurance industry has always relied on statistic to determine the insurance rates. Risk-based models form the basis for calculators that are used to calculate insurance premiums. Here is a case specific to the automotive insurance. In the United States, some of the variables in these risk-based models are reasonable but others are debatable. For example, gender is a variable that determines the insurance rate. An average American male driver pays more compared to a female driver with equivalent credentials. Today, people look upon these factors as discriminatory and demand a fairer method with higher weightage to variables that are in control of the actual drivers. The European Court of Justice has passed a ruling stating that gender cannot be used as a basis for deciding insurance premiums.[21] The current trend requires risk-based models to give consideration to individuals' statistics rather than generalized population statistics. This seems fair but does require handling significantly more data on a daily basis and new models to replace the traditional ones. Big data tools and advanced data analytics might pave the way for a fairer insurance industry of the future. Predictive analytics is also widely used by the insurance industries for fraud detection, claims analytics, and compliance & risk management.

2. **Travel & Tourism Industry:**

   The travel & tourism industry is also using big data analytics for enhancing customer experiences and offer customized recommendations. These firms use demographic statistics, average time spent by users on certain travel-related web pages, personal historic travel preferences, etc.

In order to provide better customized service, data analytics also helps the travel industry to predict when people will travel, location of traveling, purpose of traveling, etc., which can be used to assist with logistics and planning so as to provide the best customer experience at the right price. Predictive analytics is also used by travel industries for personalized offers, passenger safety, fraud detection, and predicting travel delays.

3. **Finance Industry:**

There has been a drastic or unique change seen in the financial industry in the last few years. Success in the finance industry is all about having the right information at the right time. By using big data and predictive analytics, algorithms help the industry in collecting the data from a variety of data sources and support from trading decisions to predicting default rates and risk management.

4. **Health Industry:**

The health industry produces huge amounts of data on a daily basis. The data is generated at hospitals, pharmacies, diagnostics centers, clinics, research centers, etc. Health-care industry data can have diverse data types consisting of numbers, images, x-rays, cardiograms, and even sentiments. Data analytics in health care can be used for all kinds of applications; these can include prognosis and diagnosis of an ailment, identifying the risk of propagation of a pandemic, identifying the effectiveness of a new therapy, systemic health trends in a given population, and many more. Data analytics can also be used in health care for certain non-conventional applications like tracking fraud, tracking counterfeit medicines, and optimizing patient transport logistics.

5. **Telecom Industry:**

The telecom industry has access to large amounts of customer usage and network data. By applying data analytics, it has become easier for telecom companies to understand their customer needs and behaviors in better ways and to customize the offers and services accordingly. By proving customized or personalized offers, there is a higher probability of the conversion. The telecom sector relies heavily on advance analytics for a wide variety of applications that include network optimization, fraud identification, price optimization, predicting customer churn, and enhancing the customer experience.

6. **Retail Industry:**

   The retail industry is a consumer data-driven industry where the bulk of consumer transactional data is generated on a daily basis. Data analytics is helping retailers not only in understanding customer behavior and their shopping patterns but also what they will purchase in the future. Predictive analytics is widely used by both conventional retail stores as well as e-commerce firms for analyzing their historical data and building models for customer engagement, supply chain optimization, price optimization, and space optimization and assortment planning.

7. **Agriculture Industry:**

   The agriculture industry has seen many changes in the past years and application of analytics has redefined the industry. Insights from agriculture data will help farmers in having a broader picture of their expected cost, the losses year after year, and the expected profit. It helps the agriculture industry from predicting pesticides quantities to predicting crop prices, weather conditions, soil, air quality, crop yield, and reducing waste; and the livestock health can improve the profitability for the farmers.

8. **Energy Industry:**

   Energy companies can predict the demand for energy in a particular season or time of the day and then use this to balance supply and demand of energy across various grids. The energy industry must be effective in finding out the proper balance between demand and supply flow as providing excessive energy will lead to lower profits and providing too little will make customers unhappy, find another provider, and results in customer churn. In the energy industry the data related to electricity usage, outages, transformers, and generators can also help in automated predictions, optimization of grid devices, and identifying the trend of energy usage.

Who Are Analytical Competitors?

Today all the industries are competing and are highly successful based on their analytical capabilities. Every organization is using analytics widely and methodically to outthink and outexecute the competition. Analytics is applied in the industries by the analytical competitors in order to make the smartest business decisions and come up with a distinctive capability that makes them better than anyone else in the industry. For example, Amazon's distinctive capability is "customer loyalty and better service." Figure 1-7 displays the Analytic Competitors found in every industry.

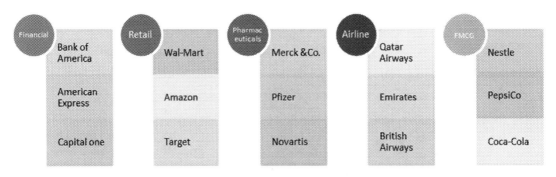

Figure 1-7. *Analytic Competitors in every industry*

Key Models and Their Applications in Various Industries

Application of data analytics in a given situation is only limited by three things: data on hand and knowledge of various models and creativity of the data scientist. In this book, six models are discussed that are used to solve a given business problem for a given industry. These models can be used to solve a wide variety of business problems across various industries. Table 1-1 shows six models discussed in this book across the six industries.

Table 1-1. Six Models Across Different Industries

Industry	Algorithms & Models					
	Logistic Regression Model	SARIMA (Time series Model)	Decision Tree	Random Forest	Multiple Linear Regression Model	RFM, K-means clustering
Banking	**Predict loan default**	Forecasting bank deposit rate[22]	Credibility of loan application[23]	Credit card fraud detection[24]	Predicting bank performance[25]	Detecting phishing of e-banking websites[26]
Retail	Company performance based on credit risk[27]	**Sales Forecasting**	Predict business failure[28]	Predict partial defection in behaviorally loyal customers[29]	Predicting energy consumption of a supermarket[30]	Grouping of retail items into fast-moving and slow-moving categories[31]
Telecom	Customer churn[32]	Forecasting the telecom income[33]	**Customer churn[34]**	Predicting fraudulent transactions[35]	Role of emotional intelligence on employees performance in customer services[36]	Mining Profitability of Telecommunication Customers[37]
Health care	Predicting patients at risk of becoming high-cost health care users	Forecasting patient arrival rates for optimal staff scheduling[38]	Predict breast cancer survivability[39]	**Predicting probability of malignant and benign cancer**	Predicting length of stay in hospital[40]	Patient segmentation for accessing patient value[41]

(*continued*)

Table 1-1. (*continued*)

Industry	Algorithms & Models						
	Logistic Regression Model	SARIMA (Time series Model)	Decision Tree	Random Forest	Multiple Linear Regression Model	RFM,	K-means clustering
Airlines	Airport satisfaction[42]	Forecast air transport demand[43]	Forecasting the no-show rate of passengers[44]	To diagnose aviation turbulence[45]	Flight Arrival delay		Passengers segmentation based on their travel behavior[46]
Fast Moving Consumer Goods (FMCG)	Customer selection[47] Customer churn[48]	Inventory & sales forecasting for short shelf-life products	Predicting customer churn	Space allocation analysis	Analyze the impact of advertisement expenses on the growth of the firm		Customer segmentation

Summary

In this chapter we have discussed about data, analytics, its types and various analytics tools used for data mining, and model building. We also discussed about the key applications of analytics that have redefined various industries. In the coming chapters, six of the major industries and the role of data analytics are discussed in detail. Each chapter provides a brief overview of the industry, lists the key firms active in that sector, and specific applications of analytics in the given industry. In each chapter a case study is presented to address a realistic business problem associated with the industry. A unique model is used to solve the given problem. The statistical background, underlying assumptions, and key features of each model are explained in detail. Codes for executing the models in both R and SAS® are provided. Various output plots and other result visualizations techniques are discussed as well as methods for easy result interpretation.

References

1. Eisenstadt, S. SAS: A hard-to-define product but simple success 1996.

2. SAS SAS company overview.

3. Studio, R. Take control of your R code.

4. Rose, R. Defining analytics: A conceptual framework ORMS Today [Online], 2016, p.

5. Cooper, D.; Schindler, P., Business Research Methods: 12th Edition. McGraw-Hill Higher Education: 2013.

6. Ball, A. Review of Data Management Lifecycle Models: 2012; p 14.

7. Russell, M., Mining the Social Web: Analyzing Data from Facebook, Twitter, LinkedIn, and Other Social Media Sites. O'Reilly Media, Incorporated: 2011.

8. Berman, J. J., Principles of Big Data: Preparing, Sharing, and Analyzing Complex Information. Elsevier Science: 2013.

9. Simon, P., Analytics: The Agile Way. Wiley: 2017.

10. Martz, E., Understanding Qualitative, Quantitative, Attribute, Discrete, and Continuous Data Types. The Minitab blog: `http://blog.minitab.com/blog/understanding-statistics/ understanding-qualitative-quantitative-attribute- discrete-and-continuous-data-types`, 2017.

11. XSI. Updated for 2017: The V's of Big Data: Velocity, Volume, Value, Variety, and Veracity 2017.

12. Harvey, C. Big Data Challenges 2017.

13. W. N. Venables, D. M. S.; Team, a. t. R. C. An Introduction to R:Notes on R: A Programming Environment for Data Analysis and Graphics V3.4.4, p. 105.

14. Wendler, T.; Gröttrup, S., Data Mining with SPSS Modeler: Theory, Exercises and Solutions. Springer International Publishing: 2016.

15. Ordenes, F. V.; Theodoulidis, B.; Burton, J.; Gruber, T.; Zaki, M., Analyzing Customer Experience Feedback Using Text Mining:A Linguistics-Based Approach. Journal of Service Research 2014, 17 (3), 278-295.

16. Sokol, D. L.; Jonas, J. Using Entity Analytics to Greatly Increase the Accuracy of Your Models Quickly and Easily.

17. Rossum, G. v.; Jr., F. L. D., The Python Language Reference Release 2.6.3. Python Software Foundation: p 115.

18. Nielsen, F. A., Data Mining with Python (Working draft). `http:// www2.compute.dtu.dk/pubdb/views/edoc_download.php/6814/ pdf/imm6814.pdf`, 2017; p 103.

19. Fabian Pedregosa, G. V., Alexandre Gramfort, Vincent Michel, Bertrand Thirion, Olivier Grisel, Mathieu Blondel, Peter Prettenhofer, Ron Weiss, Vincent Dubourg, Jake Vanderplas, Alexandre Passos, David Cournapeau, Matthieu Brucher, Matthieu Perrot, Édouard Duchesnay, Scikit-learn: Machine Learning in Python. Journal of Machine Learning Research 2011, 12, 2825-2830.

20. Sebastian Land, S. F. RapidMiner 5: RapidMiner in academic use 2012.

21. Bosari, J. What Really Goes Into Determining Your Insurance Rates? 2013.

22. Cheshti, M.; Taher Ahmadi Shadmehri, M.; Safaye Nikoo, H., Forecasting Bank Deposits Rate: Application of ARIMA and Artificial Neural Networks. 2014; Vol. 7, p 527-532.

23. He, Y.; Han, J.-c.; Zeng, S.-h., Classification Algorithm based on Improved ID3 in Bank Loan Application. 2012; p 1124-1130.

24. Seeja, K. R.; Zareapoor, M., FraudMiner: A Novel Credit Card Fraud Detection Model Based on Frequent Itemset Mining. The Scientific World Journal 2014, 2014, 10.

25. Nor Mazlina Abu Bakar, I. M. T., Applying Multiple Linear Regression and Neural Network to Predict Bank Performance. International Business research 2009, 2 (4), 8.

26. Morteza Arab, M. K. S., Proposing a new clustering method to detect phishing websites. Turkish Journal of Electrical Engineering &Computer Sciences 2017, 25, 4757-4767.

27. Hu, Y.-C.; Ansell, J., Measuring retail company performance using credit scoring techniques. European Journal of Operational Research 2007, 183 (3), 1595-1606.

28. Gepp, A.; Kumar, K.; Bhattacharya, S., Business failure prediction using decision trees. Journal of Forecasting 2010, 29 (6), 536-555.

29. Buckinx, W.; Van den Poel, D., Customer base analysis: partial defection of behaviourally loyal clients in a non-contractual FMCG retail setting. European Journal of Operational Research 2005, 164 (1), 252-268.

30. Braun, M. R.; Altan, H.; Beck, S. B. M., Using regression analysis to predict the future energy consumption of a supermarket in the UK. Applied Energy 2014, 130, 305-313.

31. Kusrini, K., Grouping of Retail Items by Using K-Means Clustering. Procedia Computer Science 2015, 72, 495-502.

32. Sebastian, H. T.; Wagh, R., Churn Analysis in Telecommunication Using Logistic Regression. Orient.J. Comp. Sci. and Technol 2017, 10 (1), 5.

33. Wang, M.; Wang, Y.; Wang, X.; Wei, Z., Forecast and Analyze the Telecom Income based on ARIMA Model. 2015; Vol. 9, p 2559-2564.

34. Kamalraj, N.; Malathi, A., A Survey on Churn Prediction Techniques in Communication Sector. 2013; Vol. 64, p 39-42.

35. Fatimah Almah Saaid, R. K., Darfiana Nur, Development of Users' Call Profiles using Unsupervised Random Forest. In Third Annual ASEARC Conference, Newcastle, Australia, 2009.

36. Chughtai, M. W.; Lateef, K., Role of Emotional Intelligence on Employees Performance in Customer Services: A Case Study of Telecom Sector of Pakistan. In International Journal of Advance Research in Computer Science and Management Studies, 2015; Vol. 3, p 8.

37. Arumawadu, H.; Rathnayaka, R. M. K.; Illangarathne, S. K., Mining Profitability of Telecommunication Customers Using K-Means Clustering. 2015; Vol. 3.

38. Ganguly, A.; Nandi, S., Using Statistical Forecasting to Optimize Staff Scheduling in Healthcare Organizations. Journal of Health Management 2016, 18 (1), 172-181.

39. Umer Khan, M.; Pill Choi, J.; Shin, H.; Kim, M., Predicting Breast Cancer Survivability Using Fuzzy Decision Trees for Personalized Healthcare. 2008; Vol. 2008, p 5148-51.

40. Combes, C.; Kadri, F.; Chaabane, S. In Predicting hospital length of stay using regression models: Application to emergency department, 10ème Conférence Francophone de Modélisation, Optimisation et Simulation- MOSIM'14, Nancy, France, 2014-11-05; Nancy, France, 2014.

41.   Wu, H.-H.; Lin, S.-Y.; Liu, C.-W., Analyzing Patient's Values by Applying Cluster Analysis and LRFM Model in a Pediatric Dental Clinic in Taiwan. The Scientific World Journal 2014, 2014, 7.

42.   Mattozo, T. C.; da Silva, G. S.; Neto, A. P. F.; Costa, J. A. F. In Logistic Regression Applied to Airport Customer Satisfaction Using Hierarchical Quality Model, Berlin, Heidelberg, Springer Berlin Heidelberg: Berlin, Heidelberg, 2012; pp 558-567.

43.   Andreoni, A.; Postorino, M. N., Time series models to forecast air transport demand: A study about a regional airport. IFAC Proceedings Volumes 2006, 39 (12), 101-106.

44.   Zenkert, D. No-show Forecast Using Passenger Booking Data. Lund University, http://lup.lub.lu.se/luur/download?func=downloadFile&recordOId=8903812&fileOId=8903813, 2017.

45.   Williams, J. K., Using random forests to diagnose aviation turbulence. Machine Learning 2014, 95 (1), 51-70.

46.   Pritscher, L.; Feyen, H., Data mining and strategic marketing in the airline industry. 2001.

47.   Tullio, I. D. Improving the direct marketing practices of FMCG retailers through better customer selection. An empirical study comparing the effectiveness of RFM (Recency, Frequency and Monetary), CHAID (Chi-squared Automatic Interaction Detection), stepwise logit (logistic regression) and ANN (Artificial Neural Networks) techniques using different data variable depths. Cranfield University, 2014.

48.   Haver, J. V. Benchmarking analytical techniques for churn modelling in a B2B context. Ghent University, https://lib.ugent.be/fulltxt/RUG01/002/351/241/RUG01-002351241_2017_0001_AC.pdf, 2017.

CHAPTER 2

Banking Case Study

The Banking Industry is one of the fundamental pillars of modern day's business. The banking sector, due to the shear nature of the industry, has always had access to vast amounts of credible consumer data. In addition to this, there are two other factors that have played a significant role in this industry being the forerunner in implementing data analytics.

The first factor is that the financial sector was quick to understand the significance of computers and was an early adopter of computer technology. In 1950 the Bank of America pioneered the adoption of computers for bookkeeping and check processing.[1] This computer, Electronics Recording Machine, Accounting (ERMA), was developed and prototyped by Stanford Research Institute (SRI) and built by the General Electric Computer Department.[2] Development of ERMA not only revolutionized bookkeeping and check processing but also brought about key changes in bank account data handling. For example, the concept of bank account numbers as a unique identifier was first conceptualized and used under this program and is used to date for handling financial data. By October of 1993 as per the U.S. Census Bureau, the private financial sector had the highest percentage of workers using computers at work.

The second factor is the regulatory environment that required the banks to have constant access to data for reporting purposes. In the early 1980s, vital changes were brought about in banking regulations and reporting to provide much needed stability to the international banking system. In 1988 the Basel Committee of Banking Regulations consisting of representatives of central banks of the Group of Ten (G10) published the "Basel Accord" or "Basel I." Basel I provided guidelines to accurately quantify the various financial risk to which the banks were vulnerable and required them to balance these risks with sufficient capital on hand at all times.[3] Basel II[4] and Basel III[5] reforms followed

Basel I to impact more stability to the banking sector. In terms of data this meant that financial institutions and banks had to have information from the balance sheets as well as off balance sheet entities to be available in an easily ***accessible and analyzable*** form for reporting purposes. These factors provided the necessary foundation for implementing data analytics by providing ***relevant credible structured data in a digital format***.

In this chapter a case study of predicting bank-loan default for the banking industry is presented. R and SAS Studio have been employed in this case study. The rest of the chapter is organized as follows. In the next section, key applications of analytics that have redefined the banking industry are discussed. The following section discusses in detail about the logistic regression model, equation, assumptions, and model fit. The section after that provides the main steps and tasks for data exploration in R. In the subsequent section, the logistic regression model is performed and model output of each part is presented in R. The section after this discusses data exploration, model building, and interpreting each part in SAS Studio. In the last section, the summary of the chapter is presented.

Applications of Analytics in the Banking Sector

Today data analytics has become almost mainstream in the banking industry, evident from the fact that 90% of top 100 banks use SAS®[6] and all of the major professional service networks including McKinsey & Company, Boston Consulting Group(BCG), Deloitte, PricewaterhouseCoopers (PwC), Ernst & Young(E&Y), Klynveld Peat Marwick Goerdeler (KPMG), and many more offer financial analytics services for the banking sector. A word cloud built based on a summary of applications provided by SAS® banking industry customers[7] is shown in Figure 2-1.

Figure 2-1. *Data analytics applications in the banking sector*

Analytics is playing an important role in redefining the banking industry in a holistic sense with applications ranging from risk mitigation, fraud detection, customer acquisition and retention, forecasting, and an increase in cross-selling & up-selling. An overview of some of these key applications is provided in this section.

Increasing Revenue by Cross-Selling and Up-Selling

Cross-selling and Up-selling are sales strategies that are applied across various industries with diverse portfolios. The modern-day banking industry is a perfect example with diverse financial offerinsg like checking and savings accounts, credit and debit cards, short-term and long-term loans, and mortgages. Cross-selling can be defined as a sales strategy to influence existing and new customers to buy complementary products or services in addition to the original purchase while Up-selling is the sales strategy to influence existing customers to purchase a higher valued product or service of the same product line.[8]

With the help of predictive analytics, modern banking is able to collect and consolidate the internal data from different departments like commercial loans, personal loans, term deposits, etc. The analytics team analyzes this data and builds models to derive insights. Creative approaches to gather customer data from social media like Twitter and to analyze the sentiments of customers by applying the sentiment analysis model or text analytics is also gaining significant popularity.

Sentiment analysis helps banks to predict the ***next product*** offerings that a customer would be interested in purchasing.[9] The effective cross-selling and up-selling is possible only when banks understand what their customers need, like are the customers happy from their products or services? Which type of investments is giving them high revenue? Did customers look for any customized or personalized products or services? And so on. Banks have various channels to cross-sell and up-sell their products and services by interacting with their customers by phone banking, Web, emails, SMS, walk-in, etc. Strong customer relationships and customer loyalty are always an added advantage in driving effective cross-selling and up-selling.[10]

Minimizing Customer Churn

Customer Churn or attrition is defined as when customers leave your service or loss of customers. Attrition in any industry hampers its growth, and it is a sign of failure for the industry.[11] Higher attrition rates signify that customers are not happy from products and service due to several reasons.

Banking analytics helps in analyzing the customers' past purchasing behavior, based on their habits, need, investments, etc., and accordingly customize their product to fulfill their need and make them happy and satisfied. A broad spectrum of predictive analytics and machine learning techniques like logistic regression,[12] decision tree, and support vector machines (SVM)[13] are used by banks to identify the probability of customer churn and identify the factors that are responsible for customer churn. Targeted strategies are developed to address factors responsible for churn as well as provide the customers with the highest risk of attrition with promotional offers and advertisements to keep them happy and satisfied. A targeted approach can help significantly reduce costs associated with customer churn and also help banks identify and address systemic problems.

Increase in Customer Acquisition

Creating or acquiring new customers is the top priority of every industry for their growth. Predictive analytics and machine learning algorithms are helping banks in creating new customers by luring them with attractive products and conducting lots of promotional and effective advertisement activities so the customers can be aware of their products and services.

Acquiring new customers always costs more than retaining old customers. Banks are deploying analytics to their complex data so they can analyze it for better insights. KYC (know your customer) helps bank to know the background of their customers in a better way, like what types of product and services customers would be interested to utilize? This will help banks to customize the products and services for the customers accordingly and keep track in managing their portfolio in order to keep them happy and satisfied as happy customers tend to purchase more products, stick forlonger time, and spreads good word of mouth about their product and services. After creating new customers, improving customer experiences is very important to retain them. To survive in the cut-throat competitive world, banks are giving high priority to improving the customer experience by applying analytical and marketing strategies.

Predicting Bank-Loan Default

Loans are the biggest source of income for the banks. Profitability for the net interest income sources like loans relies on two aspects: interest rates and low default rates.

In 2007 there was a mortgage crisis that caused bad impacts on the financial market. The main reason behind this subprime mortgage crisis[14] was heavy borrowing for home loans at lower interest rates with an assumption that house prices would inflate. Lenders used to approve loans with minimal document verification and even if the borrower lacked minimal credentials. Due to these practices, the housing market became extremely volatile and resulted ina huge number of loan defaults and foreclosures. Many U.S. investments banks went bankrupt, one famous example being Lehman Brothers.

To prevent the reoccurrence of such a financial holocaust, new banking regulations were put in place and banks implemented strict measures to minimize loan default rates.[15] These methods consisted of more stringent borrower information and document verification practices as well as developed tool sets to statistically identify a possible occurrence of default based on historic data. These tools primarily consisted of analytics and machine learning techniques like logistic regression and a decision tree to quantify the default propensity of loan applicants.

Predicting Fraudulent Activity

Bank fraud is defined as the act of using illegal ways or indulging in illegal activities in order to receive money, funds, credits, or assets from a bank or any other financial institutions.

The most common fraud threats are Phishing, credit/debit card scams, check scams, accounting fraud, identity theft, and money laundering.[16] Figure 2-2 shows a chart of different categories of fraud in the banking sector.

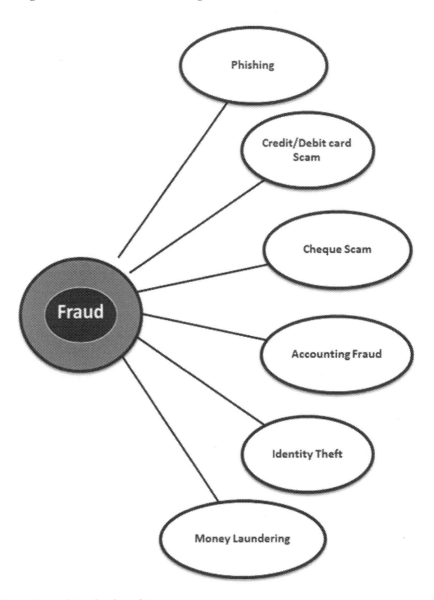

Figure 2-2. *Fraud in the banking sector*

- Phishing: In Phishing the customers will receive spam emails. Through these emails, those phishing will try to get access to the customer's account information by requesting customers to update their login and password information, credit card details, etc. Those spam emails seem to come from the bank by creating the exact website of the bank and trapping the customers in such spams. Whenever such type of emails are received, customers should ignore them and if necessary update their respective banks about such scam emails.

- Credit/Debit Card Scam: The credit/debit card fraud occurs when an individual credit card is fraudulently used by another individual or any third party to purchase goods and services, etc.

- Check Scam: Check fraud occurs when checks are altered and duplicated without the permission of the authorized person, for example, a 100-dollar check can be altered to 1000 dollars. In other cases, a lag in check encashment practices are exploited where the checks are used as forms of payments but no funds are available to support encashment.

- Accounting Fraud: Accounting fraud constitutes deliberate falsification of financial records and estimates for purposes of financial gains. Falsification of records may constitute practices like inflating assets, revenues and profit margins, and/or understating liabilities and expenses. The primary motivation of financial fraud is to lure investors or seek financial services and support from banks and other financial institutions. Various data mining, predictive analytics, and machine learning technique are gaining popularity for accounting fraud detection including outlier identification, neural networks, decision tree, and regression models.[17]

- Identity Fraud: Identity fraud in relation to the financial sector can be defined as the practice of illegally obtaining personal and financial details of an individual for the purposes of financial gains. All this personal information can be used to get the financial advantages or other benefits such as getting bank account access, loans, credit cards, etc., so the person can take out the cash and loan easily and disappear from the place, which will result in huge financial loss and victims will suffer badly.

- Money Laundering: Money laundering occurs when a large amount of money is illegally received and the true source of the money is hidden or concealed (like from terrorist activity, smuggling drugs, etc.) so that the money seems to be received from a legal source.[18]

In the money laundering process, the fraudulent money is deposited in one country and then transferred to other countries so they can use it safely. There are so many acts of money laundering, for example, the inactive bank accounts will become active suddenly and a huge amount of cash is deposited in that account, a person opening multiple accounts for doing their illegal bank transactions with the same name, and many mor . Here are some real-time cases where fraud activities in banks were responsible for huge financial loss: for example, the 1873 Bank of England Forgeries,[19] Moldovan bank fraud scandal,[20] and Russian Laundromat,[21] etc. Analytics help banks in predicting the probability of the fraudulent transactions and send the early warning signals so that preventive measures can be taken immediately and effectively, which will help both banks and their customers from huge financial loss.

Case Study: Predicting Bank-Loan Defaults with Logistic Regression Model

In the modern-day banking industry, commercial banks have diverse sources of revenue that can be broadly classified as non-interest sources and net interest sources. Non-interest sources constitute deposit account services like ATMs, online payments, safe deposits, and cash management services like payroll processing. Net interest earnings, on the one hand, are essentially comprised of different types of loan offerings where the bank makes money by lending at a higher interest rate than what it pays on deposits. In spite of non-interest incomes on rise, over half of the money made by banks still comes from net interest earnings.[22] A bank's success heavily relies upon how many loans it can give out while maintaining low default rates, where default means the inability of the borrowers to pay back the loan in time. In this case study, application of the logistic regression model to predict which customers have a high propensity to default on a bank loan is demonstrated.

Logistic regression or the logit model is defined as a type of regression model where the dependent or target variable is binary or dichotomous like having only two values as default /not default, fraud /not fraud, live/dead, etc., and independent or explanatory variables can be binary, continuous, ordinal, etc.

When a dependent or target variable has only two values, then then it is considered a binomial logistic regression.[23] In other cases when the target variable has more than two values or outcome categories, then the multinomial logistic regression is applied and if multiple outcome categories are ordered, then it is considered as ordered logit model, for example, the target variable is having multiple ordered categories like excellent, good, and average.

The logistic regression or logit model is used to model the data and describe the relationship between the target or dependent binary variable and independent or explanatory variable. The ogistic regression model is used in various fields like medical science, life science, actuarial science, and many more. For example, one needs to predict whether patients will be readmitted to a hospital or not? Whether a customer will default a bank loan or not? Will the customer churn or not? In all these cases, the target variable is binary or dichotomous; having only two values (yes or no) .Past and historical data is used to predict the future events by applying the logistic regression model when the dependent variable is binary.

Logistic Regression Equation

In logistic regression, the probability of a binary outcome is predicted; hence in a logistic regression formula, when the probability is y = 1, it is represented as P and when the probability is y = 0, it is represented as 1 – P.

$$\ln\left(\frac{P}{1-P}\right) = \beta_0 + \beta_1 x_1 + \beta_2 x_2 \cdots \cdots \beta_n x_n \tag{1}$$

Where $\ln\left(\dfrac{P}{1-P}\right)$ = Logit function

$x_1, x_2, \cdots x_n$ = Independent variables

β_0 = Logistic regression model intercept

$\beta_1, \beta_2, \cdots \beta_n$ = Logistic regression coefficients for N independent (x) variables

Odds

In logistic regression the odd ratio is defined as the probability of the occurrence of the events (1 or Yes) to the probability of the occurrence of the nonevents (0 or No). For example, there is a sample of 60 unemployed borrowers, and 50 are defaulting on the bank loan and 10 are not defaulting on thebank loan, while in another sample of 60 employed borrowers, 40 are defaulting on a bank loan and 20 are not defaulting on a bank loan.

Status	Default	Non-Default	Total Borrower
Unemployed Borrower	50	10	60
Employed Borrower	40	20	60

The Probability of defaulting on a bank loan by an unemployed borrower is calculated below.

$$P = 50/60 = 0.8$$

$$q = 1\text{-}P = 1 - 0.8 = 0.2$$

For an unemployed borrower, the probability of defaulting on a bank loan is 0.8 and the probability of not defaulting on abank loan is 0.2.

The Probability of defaulting on a bank loan by the employed borrower is calculated below.

$$P = 40/60 = 0.6$$

$$q = 1\text{-}P = 1 - 0.6 = 0.4$$

For employed persons the probability of defaulting on a bank loan is 0.6 and the probability of not defaulting on a bank loan is 0.4.

Now in the next step these probabilities are used to compute the odds of bank-loan default for unemployed and employed borrowers.

$$\text{Odds (unemployed)} = 0.8/0.2 = 4$$

$$\text{Odds (employed)} = 0.6/0.4 = 1.5$$

ODDS Ratio: ODDS Ratio is computed as the ratio of the two odds. In this example the odds ratio is computed by dividing the odds of the unemployed by odds of the employed.

$$OR = 4/1.5 = 2.66$$

Now the conclusion is that for an unemployed person. the odds of defaulting bank loans is 2.66 times higher than the odds for defaulting on bank loans for employed persons.

Logistic Regression Curve

The logistic regression curve is an S-shaped or Sigmoidal curve as shown in Figure 2-3. In a sigmoid curve the curve starts with slow linear growth, followed by exponential growth and again slows down to become stable. when dependent variable (y) is binary (0,1 or yes, no) and the independent variable (x) is numerical, the logistic regression model fits a curve to display the relationship between the dependent variable (y) and theindependent variable (x).

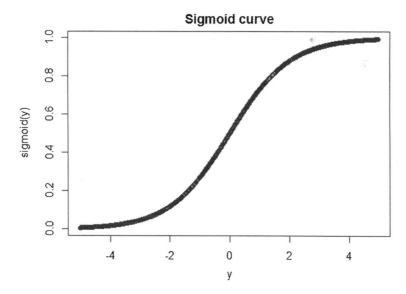

Figure 2-3. *Logistic regression sigmoidal curve*

Logistic Regression Assumptions

All statistical models are developed keeping in mind certain assumptions about the characteristics of data. It is critical that data on which the model is being applied is aligned with these assumptions for obtaining reliable results. In addition to these assumptions, there are also certain model-specific formats that need to be followed to ensure the data is aligned with the fundamental workings of the model. Some of these changes can be applied when the conditioning and structuring of the data is being performed prior to model implementation.

Key logistic regression assumptions and format are listed below:

1. In logistic regression, it is that the target variable must be discrete and mostly binary or dichotomous. This is the key differentiator between the linear regression model and the logistic model and primary requirement for the logistic model.

2. Logistic regression estimates the probability of the occurrence of events P (Y=1) so it is important that the dependent or target variable must be coded accordingly and the desired outcome must be coded to be 1.

3. Regression model should neither be overfitted nor underfitted. Overfitted data can significantly lose the predictive ability due to an erratic response to noise whereas underfitted will lack the accuracy to account for the variability in response in its entirety.[24]

4. There should not be any multicollinearity between the independent variables in the model, and all independent variables should be independent to each other.[25]

5. Independent or explanatory variables must be linearly related to the log odds, and in the logistic regression model, the linear relationship between the dependent or target variable and independent or explanatory variable is not required like as in the linear regression model.

In addition to these assumptions and format requirements, the logistic regression model requires quite a large sample size as maximum likelihood estimates (MLE) are less powerful than ordinary least square (OLS).[26]

Logistic Regression Model Fitting and Evaluation

Model fit tests are done to evaluate how well the model fits the data or how accurately the model can predict the observed values or real values. Lesser is the difference between observed values and predicted values, and better is the model. These differences between the predicted and observed values are also referred to as residuals. Most commonly used methods for evaluating the model fit are the Likelihood ratio test and Hosmer and Lemeshow test.

1. Likelihood Ratio Test: Likelihood ratio test is used to compare two nested generalized linear models. Like here, the comparison between a reduced (model with few predictors) and a full model is done.[27] The formula which is used to measure the difference of the reduced model and full model with k degrees of freedom is the following:

$$\chi^2 = -2\ln\left(of\ reduced\ model\right) - \left(-2\ln\left(likelihood\ of\ full\ model\right)\right)$$

    If p value for model fit is less than 0.05, then signify that our full model fits significantly better than our reduced model.

2. Hosmer-Lemeshow Test: Hosmer-Lemeshow test is another approach for reflecting the goodness of fit. In this test with the help of Pearson Chi-Square, it is examined whether the observed proportion of events are same to the predicted probabilities in the model population subgroup. In this test predicted probabilities are divided into deciles, which are in 10 groups.

The formula used for calculating the Hosmer- Lemeshow test is:

$$H = \sum_{G=1}^{10} \frac{\left(O_g - E_g\right)^2}{E_g}$$

Where O_g = no of observed events in the gth group.

E_g = no of expected events in the gth group.

This test follows a Chi-Square distribution with 8 degrees of freedom, which is calculated as (number of group – 2). The output of this Hosmer-Lemeshow test will be the Chi-Square value and a p value.

If the p value is small, such as the p value less than 0.05, it signifies that the model is a poor fit; and if the p value is large, with the p value greater than 0.05, it signifies that the model is a good fit. The Hosmer- \Lemeshow test is usually not recommended because of the selection of the number of subgroups.

Statistical Test for Individual Independent Variable in Logistic

Regression Model

Once the overall model fit tests are done, then the next step is to examine that how important is each independent variable or the significant contribution of the independent variables in the model.

After building the logistic regression model, rw logistic regression coefficient for an individual independent variable is known and from that we can conclude that if there is a change in 1 unit of nth independent variable, when all other independent variables are held constant, then the log odds of the predicted outcome also changes by Y_n units.

To find out the importance of an independent variable in logistic regression, there are different type of tests that can be applied like the Wald statistic test and likelihood ratio test.

1. Wald Statistic test: The Wald statistic test is used to find out the importance of the independent variables in a model or how efficient and effective are independent variables in predicting the dependent variable by looking at the logistic regression coefficient of individual independent variables.

    The Wald statistic test is calculated by using this formula:[28]

$$W_J = \left(\frac{B_J}{SE \ of \ B_J} \right)^2$$

    Where:

    B_J - is the regression coefficient.
    $SE \ of \ B_J$ - is the standard error of the regression coefficient.

In Wald statistics, the null hypothesis is when the coefficient of interest is equal to zero and the alternate hypothesis is when coefficients are not equal to zero. If the Wald statistic test accepts the null hypothesis, then we can conclude that removing that variable from the model will not impact the model fit.

Wald statistic test is easy to calculate but it has drawbacks, in some cases when data has large coefficients estimates, which tends to increase the standard error and hence lower the value of the Wald statistic. Due to the lower value of the Wald statistic the independent variables can be considered insignificant in the model, though they are playing significant roles in the model.

2. Likelihood ratio test: In addition to evaluation of the model fit, the likelihood ratio test is also used to find out the contribution of individual independent variables in the model. This test is calculated by using the formula:

$$G = -2\big(\ln\big(Reduced\ model\big) - \ln\big(Full\ model\big)\big)$$

$$\chi^2 = -2\ln\big(of\ reduced\ model\big) - \big(-2\ln\big(likelihood\ of\ full\ model\big)\big)$$

The individual independent variable is entered inside the model in an orderly manner, and then the comparison between both the models is done in order to study the contribution of each independent variable in the model. The smaller the deviance between the reduced model and the full model, the better is the correlation between the dependent or target and independent or explanatory variables.

Predictive Value Validation in Logistic Regression Model

In predictive value validation, a few measurements like the confusion matrix and receiver operating characteristic are used to find out that how accurately the model is predicting the target or dependent variable or, in other words, we can say it measures the model accuracy.

- Confusion Matrix:

   The confusion matrix is a technique used to evaluate the
   predictive accuracy or the performance of the logistic regression
   model; it is a classification table with two dimensions consisting of
   2 rows and 2 columns. Each column represents the dichotomous
   or binary predicted class and each row represents the
   dichotomous or binary actual class.[29]

Let's take an example to explain the confusion matrix for a logistic regression model when the dependent variable is binary or dichotomous. In this example, if we were predicting customer churn, Yes means the customers would churn and No means the customers would not churn. There are a total of 185 customers that were being tested for customer churn. Out of 185 events, the classifier predicted 130 times Yes and 55 times No. In Actual, there are total 125 customers in the sample data who are churning and 60 customers that do not churn.

A sample classification table is provided in Table 2-1. Columns are representing binary or dichotomous predicted outcomes and rows are representing binary or dichotomous actual outcomes.

Table 2-1. *Classification Table*

N = 185		Predicted Outcomes	
Actual Outcomes	**NO**	**YES**	**Total Rows**
NO	50 (TN)	10 (FP)	**60**
YES	5 (FN)	120 (TP)	**125**
Total Columns:	**55**	**130**	**185**

In the above confusion matrix table there are True Positive, True Negative, False Positive, and False Negative cases displayed.

- True Positive (TP): In this case we predicted yes (customers would churn, and in reality they do churn.

- True Negative (TN): In this case we predicted no (customers would not churn), and in reality they don't churn.

- False Positive (FP): In this case we predicted yes (customers would churn) but in reality they don't churn; this is also known as a type l error.

- False Negative (FN): In this case we predicted no (customers would not churn) but in reality they do churn, and this is also known as a type ll error.

  Binary classifier confusion matrix is used for computing various rates like accuracy rate, error rate, true positive rate, false positive rate, specificity, precision, and prevalence. These terms are defined in the following section.

- Accuracy Rate: Accuracy rate is the total fraction of accurately predicted outcomes to the total outcomes. It is calculated by the formula:

$$\frac{TP + TN}{TP + FP + FN + TN}$$

$$\frac{120 + 50}{50 + 10 + 5 + 120} = \frac{170}{185} = 0.92$$

  Where TP = True positive
  TN = True negative
  FP = False positive
  FN = False negative

- Error Rate: Error rate is also known as misclassification rate, and it is the total fraction of inaccurately predicted outcomes to the total outcomes. It is calculated as

$$1 - Accuracy\ rate$$

$$1 - 0.91 = 0.09$$

- True Positive Rate: TPR is defined as the total fraction of accurately predicted positive outcomes to the total actual positive outcomes. It is also known as sensitivity. It is calculated by the formula

$$\frac{TP}{TP + FN} = \frac{120}{125} = 0.96$$

- False Positive Rate: False positive rate (FPR) is defined as the fraction of inaccurately predicted positive outcomes to the actual negative outcomes, which in simpler terms means how frequently does the model predict yes, when in reality it is no. It is calculated by the formula

$$\frac{FP}{TN + FP} = \frac{10}{60} = 0.17$$

- Specificity: Defined as the fraction of accurately predicted negative outcomes to the actual negative outcomes, which in simpler terms means how frequently does the model predict no when in reality it is no. It is calculated by the formula

$$1 - FPR = 1 - 0.17 = 0.83$$

- Precision: Precision is defined as the fraction of accurately predicted positive outcomes to the total positive outcomes or in simpler words how frequently is the model correct, when it predicts yes. The formula for precision is

$$\frac{TP}{TP + FP} = \frac{120}{130} = 0.92$$

- Prevalence: Prevalence is defined as the fraction of actual positive outcomes to the total outcomes. it is calculated by the formula:

$$\frac{TP + FN}{TN + TP + FP + FN} = \frac{125}{185} = 0.68$$

- Receiver Operating Characteristics (ROC) and Area Under Curve(AUC): ROC curve is a graph with the True positive rate (TPR) or Sensitivity plotted on the y-axis against the False positive rate (FPR) or 1 – Specificity on the x-axis. TPR, in simple words, is how frequently does the model predict yes when in reality it is yes, and FPR is how frequently does the model predict yes, when in reality it is no.

The ROC curve is used to measure the binary or dichotomous classifier performance visually and Area Under Curve (AUC) is used to quantify the model performance. Generally an AUC greater than 70% is considered as the accurate model. The more is the AUC curve toward 1, the better is the predictive accuracy of the model.

Receiver Operating Characteristic (ROC) and Area Under Curve is shown in Figure 2-4 and Figure 2-5. The curve above the diagonal represents the better performance, and below the diagonal it represents the worse performance and approaching toward 1, it represents the best performance.

The Area Under Curve (AUC) values varies from range 0.5 to 1.0 where 0.5 means the poor predictive ability and 1.0 is considered as the accurate or best predictive ability. The graph approaching toward 1 signifies that the model predictive accuracy is high. The higher is the AUC value, the better is the model predictive accuracy; and the lower is the AUC value, the worse is the model predictive accuracy.

Figure 2-4 and Figure 2-5 shows the ROC and AUC curve for Model A and Model B respectively. Model A (AUC) value is 68%and Model B (AUC) value is 97%. Hence Model B is having high predictive accuracy compared to Model A.

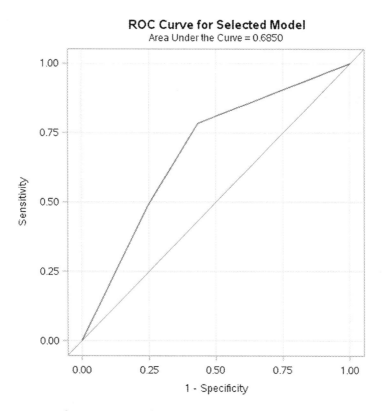

Figure 2-4. *ROC and AUC Curve for Logistic regression: Model A*

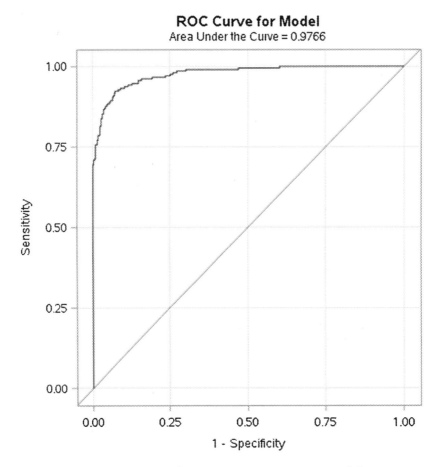

Figure 2-5. *ROC and AUC Curve for Logistic regression: Model B*

Logistic Regression Model Using R

In this banking case study, we will discuss about the data and the variables used in the data. We then discuss about the exploratory data analysis in R, which is considered as the first step in the data analysis process. Exploratory data analysis helps in taking a broad look at patterns, trends, summary, outliers, missing values, and so on in the existing data. We also discuss about building a logistic regression model and interpretation of its output in R.

Business Problem: To predict the probability of the bank-loan default.

Business Solution: To build the logistic regression model.

About Data

In this banking case study, in order to illustrate the bank-loan default, logistic regression model is created and the data is generated synthetically. In this dataset there are a total of 1,000 observations and 16 variables; 13 variables are numeric and 3 variables are categorical.

The loan-default data set contains information on 1,000 loan applicants. Default is the dependent or target variable in the data, where 1 denotes presence of loan default and 0 denotes absence of loan default. There are 70% of non-defaulter applicants and 30% applicants are loan defaulters in this data . This data set is used to develop the logistic regression model in order to predict the probability of the bank-loan default.

In R, create your own working directory to import the dataset.

```
#Read the data from the working directory, create your own working directly
to read the dataset.

setwd("C:/Users/Deep/Desktop/data")

data1 <- read.csv ("C:/Users/Deep/Desktop/data/
loan_default.csv",header=TRUE,sep=",")

data2<-data.frame(data1)
```

Performing Data Exploration

In Exploratory data analysis, we are going to take a broad look at patterns, trends, summary, outliers, missing values, and so on in the existing data. R codes for data exploration and their output is discussed in the section below.

```
#perform exploratory data analysis to know about the data

# display top 6 rows of dataset to see how data look like

head (data2)
```

	Default	Checking_amount	Term	Credit_score	Gender	Marital_status	Car_loan
1	0	988	15	796	Female	Single	1
2	0	458	15	813	Female	Single	1
3	0	158	14	756	Female	Single	0
4	1	300	25	737	Female	Single	0
5	1	63	24	662	Female	Single	0
6	0	1071	20	828	Male	Married	1

	Personal_loan	Home_loan	Education_loan	Emp_status	Amount	Saving_amount
1	0	0	0	employed	1536	3455
2	0	0	0	employed	947	3600
3	1	0	0	employed	1678	3093
4	0	0	1	employed	1804	2449
5	0	0	1	unemployed	1184	2867
6	0	0	0	employed	475	3282

	Emp_duration	Age	No_of_credit_acc
1	12	38	1
2	25	36	1
3	43	34	1
4	0	29	1
5	4	30	1
6	12	32	2

```
# display bottom 6 rows to see what data looks like

tail(data2)
```

	Default	Checking_amount	Term	Credit_score	Gender	Marital_status
995	0	589	20	733	Male	Married
996	1	17	21	562	Male	Married
997	0	590	18	873	Female	Single
998	0	343	16	824	Male	Married
999	0	709	16	811	Male	Married
1000	0	991	15	794	Male	Married

	Car_loan	Personal_loan	Home_loan	Education_loan	Emp_status	Amount
995	1	0	0	0	unemployed	829
996	1	0	0	0	unemployed	1328
997	1	0	0	0	unemployed	1433
998	0	1	0	0	unemployed	1465
999	1	0	0	0	unemployed	1359
1000	0	1	0	0	unemployed	1321

	Saving_amount	Emp_duration	Age	No_of_credit_acc
995	3171	70	33	4
996	2658	118	28	5
997	3469	108	29	5
998	3426	99	31	4
999	3114	113	28	6
1000	3309	95	33	8

describe the structure of data , it displays the datatype of each
variable present in the data like whether that particular varibale is
numeric , factor etc.

```
str(data2)
```

```
'data.frame'       : 1000 obs. of  16 variables:
 $ Default         : int  0 0 0 1 1 0 0 0 0 1 ...
 $ Checking_amount : int  988 458 158 300 63 1071 -192 172 585 189 ...
 $ Term            : int  15 15 14 25 24 20 13 16 20 19 ...
 $ Credit_score    : int  796 813 756 737 662 828 856 763 778 649 ...
 $ Gender          : Factor w/ 2 levels "Female","Male": 1 1 1 1 1 2 2
                     1 1 2 ...
 $ Marital_status  : Factor w/ 2 levels "Married","Single": 2 2 2 2 2 1
                     2 2 2 1 ...
 $ Car_loan        : int  1 1 0 0 0 1 1 1 1 1 ...
 $ Personal_loan   : int  0 0 1 0 0 0 0 0 0 0 ...
 $ Home_loan       : int  0 0 0 0 0 0 0 0 0 0 ...
 $ Education_loan  : int  0 0 0 1 1 0 0 0 0 0 ...
 $ Emp_status      : Factor w/ 2 levels "employed","unemployed": 1 1 1 1
                     2 1 1 1 2 1 ...
```

```
$ Amount          : int  1536 947 1678 1804 1184 475 626 1224 1162 786 ...
$ Saving_amount   : int  3455 3600 3093 2449 2867 3282 3398 3022 3475
                         2711 ...
$ Emp_duration    : int  12 25 43 0 4 12 11 12 12 0 ...
$ Age             : int  38 36 34 29 30 32 38 36 36 29 ...
$ No_of_credit_acc: int  1 1 1 1 1 2 1 1 1 1 ...
```

```
#display the column name of the data

names(data2)
```

```
 [1] "Default"         "Checking_amount"  "Term"
 [4] "Credit_score"    "Gender"           "Marital_status"
 [7] "Car_loan"        "Personal_loan"    "Home_loan"
[10] "Education_loan"  "Emp_status"       "Amount"
[13] "Saving_amount"   "Emp_duration"     "Age"
[16] "No_of_credit_acc"
```

```
#display the summary or descriptive statistics of the data

summary(data2$Amount)
```

```
   Min.  1st Qu.  Median   Mean  3rd Qu.    Max.
    244     1016    1226   1219     1420    2362
# Check the missing values present in the data
sum(is.na(data2))
[1] 0
is.na(data2)
```

```
      Default Checking_amount  Term Credit_score Gender Marital_status
[1,]    FALSE           FALSE FALSE        FALSE  FALSE          FALSE
[2,]    FALSE           FALSE FALSE        FALSE  FALSE          FALSE
[3,]    FALSE           FALSE FALSE        FALSE  FALSE          FALSE
[4,]    FALSE           FALSE FALSE        FALSE  FALSE          FALSE
[5,]    FALSE           FALSE FALSE        FALSE  FALSE          FALSE
[6,]    FALSE           FALSE FALSE        FALSE  FALSE          FALSE
```

	Car_loan	Personal_loan	Home_loan	Education_loan	Emp_status	Amount
[1,]	FALSE	FALSE	FALSE	FALSE	FALSE	FALSE
[2,]	FALSE	FALSE	FALSE	FALSE	FALSE	FALSE
[3,]	FALSE	FALSE	FALSE	FALSE	FALSE	FALSE
[4,]	FALSE	FALSE	FALSE	FALSE	FALSE	FALSE
[5,]	FALSE	FALSE	FALSE	FALSE	FALSE	FALSE
[6,]	FALSE	FALSE	FALSE	FALSE	FALSE	FALSE

	Saving_amount	Emp_duration	Age	No_of_credit_acc
[1,]	FALSE	FALSE	FALSE	FALSE
[2,]	FALSE	FALSE	FALSE	FALSE
[3,]	FALSE	FALSE	FALSE	FALSE
[4,]	FALSE	FALSE	FALSE	FALSE
[5,]	FALSE	FALSE	FALSE	FALSE
[6,]	FALSE	FALSE	FALSE	FALSE

In this case partial output is displayed. False represents that there are no missing values present in the data, and if there are missing values present in any variables in the data, it will be represented as TRUE.

If there is a large number of missing values present in the data, then the accuracy of the model is impacted. Model accuracy is impacted by missing values as it reduces statistical power of a study and can produce a biased estimate, leading to incorrect conclusions. To have an effective and accurate model, it is necessary to impute missing values; there are so many missing value imputation techniques like mean, K- nearest neighbor, fuzzy K-means, and many more that can be applied to handle the missing values present in the data.[30]

```
#To find out the correlation between the variables
corr <- cor.test(data2$Default, data2$Term,
method = "pearson" )
corr

    Pearson's product-moment correlation

data:  data2$Default and data2$Term
t = 11.494, df = 998, p-value < 2.2e-16
```

alternative hypothesis: true correlation is not equal to 0
95 percent confidence interval:
 0.2859874 0.3955278

sample estimates:
      cor
0.3419185

The Pearson moment correlation method is used to find out the correlation between the two variables. There is moderate positive correlation (34%) between the default and term variable; it means both of the variables are directly proportional to each other. Default increases with the increase in term. To determine whether the correlation between variables is significant, we need to compare th p value to the significance level (0.05). In this case the p- value for the correlation between default and term is less than the significance level of 0.05, which indicates that the correlation coefficient is significant. Correlation is generally of three types: positive correlation, negative correlation, and neutral correlation. When both the variables are directly proportional to each other, then there is positive correlation between the variables; when both the variables are indirectly proportional to each other, then there is negative correlation between the variables; and when there is no relationship between the two variables, then there is neutral correlation.[31]

Model Building and Interpretation of Full Data

In this case, initially we are using complete data for model building, and interpretation and in the next section, we are randomly splitting data into two parts: train data set and test data set. Train data set is used to build the model and test data set is used to test the performance of the model. There are variations seen in the output of both of the models due to the sampling. A detailed description is seen in the section below.

```
#building logistic regression model using glm on full data
```

```
fullmodel1 <-glm(Default~.,data = data2,family=binomial
(link=logit ))
```

```
summary(fullmodel1)
```

```
Call:
glm(formula = Default ~ ., family = binomial(link = logit),
data = data2)

Deviance Residuals:
    Min      1Q    Median      3Q      Max
 -3.1820  -0.1761  -0.0439   0.0415   3.3985
```

Looking at the coefficients in Listing 2-1, there are variables with p values less than 0.05; these variables are the significant variables in the model. Other variables like Gender, Marital_status, Car_loan, Personal_loan, Home_loan, Education_loan, Emp_status, Amount, Emp_duration, and No_of_credit_acc where p values are greater than 0.05, are not significant. Nonsignificant variables are removed from the model and the model is rerun. See Listing 2-1.

Listing 2-1. Coeffiecient Table

```
Call:
glm(formula = Default ~ ., family = binomial(link = logit), data = data2)

Deviance Residuals:
    Min      1Q    Median      3Q      Max
 -3.1820  -0.1761  -0.0439   0.0415   3.3985
```

Coefficients:

	Estimate	Std. Error	z value	Pr(>|z|)	
(Intercept)	39.6415229	4.7284136	8.384	< 2e-16	***
Checking_amount	-0.0050880	0.0006759	-7.528	5.14e-14	***
Term	0.1703676	0.0520728	3.272	0.00107	**
Credit_score	-0.0109793	0.0020746	-5.292	1.21e-07	***
GenderMale	0.1950806	0.5095698	0.383	0.70184	
Marital_statusSingle	0.3351480	0.4920120	0.681	0.49576	
Car_loan	-0.6004643	2.7585197	-0.218	0.82768	
Personal_loan	-1.5540876	2.7585124	-0.563	0.57318	
Home_loan	-3.5684378	2.8457131	-1.254	0.20985	
Education_loan	0.6498873	2.7894965	0.233	0.81578	
Emp_statusunemployed	0.5872532	0.3474376	1.690	0.09098	.
Amount	0.0008026	0.0005114	1.569	0.11653	

```
Saving_amount           -0.0048212  0.0006085  -7.922 2.33e-15 ***
Emp_duration             0.0029178  0.0044391   0.657  0.51099
Age                     -0.6475369  0.0646616 -10.014  < 2e-16 ***
No_of_credit_acc        -0.0968614  0.1006467  -0.962  0.33585
---
Signif. codes:  0 '***' 0.001 '**' 0.01 '*' 0.05 '.' 0.1 ' ' 1

(Dispersion parameter for binomial family taken to be 1)

    Null deviance: 1221.73  on 999   degrees of freedom
Residual deviance:  297.65  on 984   degrees of freedom
AIC: 329.65

Number of Fisher Scoring iterations: 7

# removing insignificant variables inorder to build final logistic model on
full data

fullmodel2<-glm(Default~Checking_amount+Term+Credit_score
+Saving_amount+Age,data = data2,family=binomial(link=logit))

summary(fullmodel2)
```

Now rebuilding the model again with only selected significant variables and looking at the coefficients, in Listing 2-2, the p values less than 0.05 are considered significant variables in the model.

Listing 2-2. Coeffiecient Table

```
Call:
glm(formula = Default ~ Checking_amount + Term + Credit_score +
Saving_amount + Age,
family = binomial(link = logit), data = data2)
Deviance Residuals:
    Min       1Q   Median       3Q      Max
-2.9474  -0.2083  -0.0548   0.0696   3.3564
```

```
Coefficients:
                   Estimate   Std. Error   z value   Pr(>|z|)
(Intercept)       38.8480502    3.5109294    11.065   < 2e-16  ***
Checking_amount   -0.0048409    0.0006180    -7.834   4.74e-15 ***
Term               0.1748115    0.0473527     3.692   0.000223 ***
Credit_score      -0.0113945    0.0019752    -5.769   7.99e-09 ***
Saving_amount     -0.0045122    0.0005515    -8.182   2.80e-16 ***
Age               -0.6285817    0.0587648   -10.697   < 2e-16  ***
---
Signif. codes:  0 '***' 0.001 '**' 0.01 '*' 0.05 '.' 0.1 ' ' 1

(Dispersion parameter for binomial family taken to be 1)

    Null deviance: 1221.73  on 999  degrees of freedom
Residual deviance:  339.97  on 994  degrees of freedom
AIC: 351.97

Number of Fisher Scoring iterations: 7
```

Listing 2-2 shows the coefficients (labeled Estimates), their standard errors, the Z values, and the P values. The coefficients for Checking_amount, Term, Credit_score, Saving_amount, and Age are statistically significant as their P values are less than 0.05. In the logistic regression model for a one-unit increase in the predictor variables, it will display the change (increase or decrease) in the log odds of the outcome. It can be explained as:

1. For every one-unit change in **Checking_amount**, the log odds of bank-loan default versus bank- loan non-default decreases by **(-0.004).** It signifies that the probability of bank-loan default decreases with the increase in the Checking_amount. Higher is the Checking_amount, lesser is the probability of bank-loan default and lower is the Checking_amount, the higher is the probability of bank-loan default.

2. Similarly it is for **Term** that one-unit change in Term, the log odds of bank-loan default versus bank-loan non-default increases by *(0.174).* It signifies that the probability of bank-loan default increases with the increase in the Term. The greater the Term, the higher is the probability of bank-loan default; and the less is the Term, the lower is the probability of bank-loan default.

```
#splitting data set into training and validation dataset
in 70:30
```

Now the data is split into two parts, train data set and test data set; the splitting ratio is 70:30, which means 70& of data contribute to the train dataset and 30% percent of data contribute to the test dataset. Train dataset is used to build the model and test dataset is used to test the performance of the model.

```
train_obs <- floor (0.7*nrow (data2))
print(train_obs)

#Set seed in order to reproduce the sample

set.seed(2)
train_ind <- sample(seq_len(nrow(data2)),size=train_obs)
test <-  -train_ind

#No of observations in train dataset
train_data<-data2[train_ind,]

# No of observations  in test dataset

test_data<-data2[-train_ind,]
```

Model Building and Interpretation of Training and Testing Data

```
#Building logistic regression model using glm on training data

model1<-glm(Default~.,data= train_data,family=binomial(link=logit))

summary(model1)
```

```
Call:
glm(formula = Default ~ ., family = binomial(link = logit),
data = train_data)

Deviance Residuals:
    Min        1Q     Median        3Q        Max
-2.44674   -0.13513   -0.03275   0.02618   3.05506
```

Looking at the coefficients in Listing 2-3, there are variables whose p values are less than 0.05; hence all those variables are the significant variables in the model, but there are few variables like Gender, Marital_status, Car_loan, Personal_loan, Home_loan, Education_loan, Amount, Emp_duration, and No_of_credit_acc whose p values are not less than 0.05; hence all these variables are not significant variables in the model.

Listing 2-3. Coeffiecient Table

```
Call:
glm(formula = Default ~ ., family = binomial(link = logit),
data = train_data)

Deviance Residuals:
    Min        1Q     Median        3Q        Max
-2.44674   -0.13513   -0.03275   0.02618   3.05506

Coefficients:
                        Estimate Std. Error z value Pr(>|z|)
(Intercept)           39.9012820 10.6353095   3.752 0.000176 ***
Checking_amount       -0.0058119  0.0009446  -6.152 7.63e-10 ***
Term                   0.2332288  0.0709242   3.288 0.001008 **
Credit_score          -0.0125048  0.0029325  -4.264 2.01e-05 ***
GenderMale             0.6522404  0.6859874   0.951 0.341703
Marital_statusSingle   0.4063002  0.6492109   0.626 0.531422
Car_loan               1.8317824  9.5329269   0.192 0.847622
Personal_loan          0.7887203  9.5337033   0.083 0.934066
Home_loan             -1.3617547  9.5933313  -0.142 0.887121
Education_loan         3.4385998  9.5529812   0.360 0.718884
Emp_statusunemployed   1.0376664  0.4708135   2.204 0.027525 *
Amount                 0.0005337  0.0006388   0.835 0.403508
```

```
Saving_amount          -0.0050924  0.0007743  -6.577 4.82e-11 ***
Emp_duration            0.0042426  0.0058951   0.720 0.471728
Age                    -0.7256567  0.0904776  -8.020 1.05e-15 ***
No_of_credit_acc       -0.0165636  0.1232871  -0.134 0.893126
---
Signif. codes:  0 '***' 0.001 '**' 0.01 '*' 0.05 '.' 0.1 ' ' 1

(Dispersion parameter for binomial family taken to be 1)

    Null deviance: 846.57  on 699  degrees of freedom
Residual deviance: 184.38  on 684  degrees of freedom
AIC: 216.38

Number of Fisher Scoring iterations: 8
```

Let's knock out all these insignificant variables from the model and run the model again.

```
#After removing insignificant variable inorder to build final logistic
model on training data

model2<-glm(Default~Checking_amount+Term+Credit_score
+Emp_status+Saving_amount+Age,data= train_data,family=binomial(link=logit))

summary(model2)

Call:
glm(formula = Default ~ Checking_amount + Term + Credit_score + Emp_status
+ Saving_amount + Age, family = binomial
(link = logit), data = train_data)

Deviance Residuals:
    Min       1Q   Median       3Q      Max
-2.1098  -0.1742  -0.0394   0.0343   3.2053
```

Listing 2-4 shows the coefficients (labeled Estimates), their standard errors, the Z values, and the P values. The coefficients for Checking_amount, Term, Credit_score, Emp_status, Saving_amount, and Age are statistically significant as their P values are less than 0.05.

Listing 2-4. Coefficients Table

```
Call:
glm(formula = Default ~ Checking_amount + Term + Credit_score +
Emp_status + Saving_amount + Age,
family = binomial(link = logit), data = train_data)

Deviance Residuals:
    Min       1Q   Median       3Q      Max
-2.1098  -0.1742  -0.0394   0.0343   3.2053

Coefficients:
                         Estimate Std. Error z value Pr(>|z|)
(Intercept)            43.3709264  4.9060101   8.840  < 2e-16 ***
Checking_amount        -0.0056965  0.0008912  -6.392 1.64e-10 ***
Term                    0.2077360  0.0623154   3.334 0.000857 ***
Credit_score           -0.0147534  0.0028522  -5.173 2.31e-07 ***
Emp_statusunemployed    0.9124408  0.3942729   2.314 0.020655 *
Saving_amount          -0.0049131  0.0007260  -6.768 1.31e-11 ***
Age                    -0.6885308  0.0806182  -8.541  < 2e-16 ***
---
Signif. codes:  0 '***' 0.001 '**' 0.01 '*' 0.05 '.' 0.1 ' ' 1

(Dispersion parameter for binomial family taken to be 1)

    Null deviance: 846.57  on 699  degrees of freedom
Residual deviance: 210.78  on 693  degrees of freedom
AIC: 224.78

Number of Fisher Scoring iterations: 8
```

In the logistic regression model, for a one-unit increase in the predictor variables, it will display the change (increase or decrease) in the log odds of the outcome. It can be explained as:

1. For every one-unit change in **Checking_amount**, the log odds of bank-loan default versus bank-loan non-default decreases by **(-0.005).**

2. Similarly it is for **Term** that one-unit change in Term, the log odds of bank-loan default versus bank-loan non-default increases by *(0.207).*

The coefficient for the categories of Emp_status interpretation is a bit different.

1. For example, defaulting on a bank loan with a Emp_
   statusunemployed versus with a Emp_statusemployed (omitted
   category) increases the log odds of the bank-loan default by (0.912).
   It signifies that the probability of bank-loan default increases
   with the increase in the Emp_statusunemployed versus Emp_
   statusemployed, which is considered here as the reference group,
   hence omitted category. More is the Emp_statusunemployed, the
   higher is the probability of bank-loan default.

   ```
   #Check for variance inflation factor, VIF > 5 to 10- high
   correlation
   ```

Variance Inflation Factor is a test for multicollinearity. VIF test helps in identifying the multicollinearity between the independent variables present in the data.

```
#install car package
install.packages("car")
library(car)

vif(model2)
```

Variance Inflation Factor is a test for multicollinearity,[32] and this VIF test helps in identifying the multicollinearity between the independent variables present in the data. VIF values higher than 5 to 10 display that there is multicollinearity present and the variables having high multicollinearity values is dropped from the model building, so there is no impact in the accuracy of the model. In Listing 2-5, none of the variables is shown as having VIF values higher than 5 to 10; hence no multicollineraity present in the model.

Listing 2-5. VIF Table

	VIF
Checking_amount	1.202188
Term	1.035386
Credit_score	1.096243
Emp_status	1.095911
Saving_amount	1.186483
Age	1.306097

After applying the predict function in order to predict the probability of bank-loan default and setting the probability cut-off value, like probability greater than 70%, display it as 1 (bank-loan default) else 0 (bank-loan non-default). It means all predicted as 1 or probability greater than 70 percent are considered as high likelihood of defaulting a bank loan.

```
# Predicting the model using test data

Prob <-predict(model2,test_data,type ="response")

prob1<- data.frameProb

# setting the cutoff for probability values

results <- ifelse(prob1 > 0.7,1,0)
```

Predictive Value Validation

In predictive value validation, a few measurements like the confusion matrix and receiver operating characteristic are used to find out that how accurately the model is predicting the target or dependent variable, or in other words we can say it measures the model accuracy. A detailed explanation of each concept is already explained in the above predictive value validation in the logistic regression model section.

```
#Display the confusion matrix or classification table

table(testing_high,results)
#Calculating the error rate

misclasificationerror <- mean(results != testing_high)
misclasificationerror

# Calculating the accuracy rate

accuracyrate <- 1-misclasificationerror
print(accuracyrate)
```

In the classification table shown in Table 2-2, testing_high is considered as the actual outcomes and results is considered as the predicted outcomes.

Table 2-2. *Classification Table*

	results	
testing_high	0	1
0	197	8
1	17	78

Diagonal values are correctly classified; hence that the accuracy rate can be calculated as:

$$Accuracy\ rate = \frac{197 + 78}{197 + 8 + 17 + 78} = \frac{275}{300} = 0.91$$

Error rate or Misclassification error rate can be calculated as:

$$1 - Accuracy\ rate = 1 - 0.91 = 0.09$$

```
#Calculating the confusion matrix and statistics
install.packages("caret")
library(caret)
confusionMatrix(prediction, testing_high,positive ="1")
```

In the confusionMatrix if there are only two factor levels in the data then the first level by default is used as the "positive" result. In this data that is not the case as the first level, 0 is negative so to here we override the default and correctly assign positive as "1".

```
Confusion Matrix and Statistics

          Reference
Prediction   0   1
         0 197  17
         1   8  78
```

```
            Accuracy : 0.9167
              95% CI : (0.8794, 0.9453)
 No Information Rate : 0.6833
 P-Value [Acc > NIR] : <2e-16

               Kappa : 0.8024
 Mcnemar's Test P-Value : 0.1096

         Sensitivity : 0.8211
         Specificity : 0.9610
      Pos Pred Value : 0.9070
      Neg Pred Value : 0.9206
          Prevalence : 0.3167
      Detection Rate : 0.2600
Detection Prevalence : 0.2867
   Balanced Accuracy : 0.8910

      'Positive' Class : 1
```

A detailed explanation of the statistic value like sensitivity, specificity, etc., is explained in detail in the predictive value validation in logistic regression model section.

```
# conducting Receiver operating characteristic (ROC) test and Area under curve (AUC)

install.packages("ROCR")

library(ROCR)

# Compute AUC for predicting Default with the model

prob <- predict(model2, newdata=test_data, type="response")

pred <- prediction(prob, test_data$Default)
```

```
pmf <- performance(pred, measure = "tpr", x.measure = "fpr")

plot(pmf,col= "red" )

auc <- performance(pred, measure = "auc")

auc <- auc@y.values[[1]]
auc
```
[1] **0.9688318**

Receiver Operating Characteristic (ROC) and Area Under Curve is displayed in Figure 2-6. In the y-axis true positive rate (TPR) or sensitivity is displayed and x axis false positive rate (FPR) or 1 – specificity is displayed, the curve approaching toward 1 represents the best performance of the model.

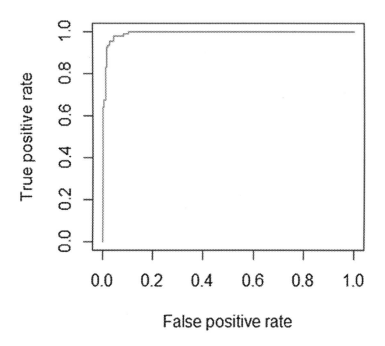

Figure 2-6. *ROC & AUC curve*

Area under Curve (AUC) values greater than 70% are considered as the model with high predictive accuracy. In this case study illustrating how the logistic regression model, Area under curve (AUC) value is around 96%, which is considered an accurate model.

Logistic Regression Model Using SAS

In this section, we discuss about different SAS procedures like proc content; proc means proc freq and proc univariate. We also discuss about building logistic regression model for predicting bank-loan default on complete data, with explanation of SAS codes and output of each part in program 1. And in the next section we discuss about splitting the data set into two parts: train data set and test data set with explanation of SAS codes and output of each part in program 2, 2.1, and section 2.2.

```
/* Create your own library in SAS like here it is libref and mention the
path */
libname libref "/home/aro1260/deep";

/* Importing loan default dataset */
PROC IMPORT DATAFILE= "/home/aroragaurav1260/data/loan_default.csv"
     DBMS=CSV Replace
        OUT=libref.loan_default;
        GETNAMES=YES;
RUN;

/* To check the no of missing values in numeric var in data */

proc means data = libref.loan_default NMISS N ;
run;
```

To check the number of missing values in numeric variables in data, N Miss represents the no of missing obs and N represents no of non-missing obs in data. In the below output of the Means Procedure shown in Table 2-3, the no of N Miss is 0 for each variable; hence no missing values present in the data.

Table 2-3.

The MEANS Procedure

Variable	N Miss	N
Default	0	1000
Checking_amount	0	1000
Term	0	1000
Credit_score	0	1000
Car_loan	0	1000
Personal_loan	0	1000
Home_loan	0	1000
Education_loan	0	1000
Amount	0	1000
Saving_amount	0	1000
Emp_duration	0	1000
Age	0	1000
No_of_credit_acc	0	1000

```
/* To check the contents of the data */
PROC CONTENTS DATA=libref.loan_default;
```

RUN; In this case partial output of procedure content is displayed in Table 2-4. It displays the content of the data like the numberof observations, number of of variables in the data, library, name and data type of each variable whether the variables are numeric or character with their Length, Format, or Informat.

Table 2-4.

The CONTENTS Procedure

Data Set Name	LIBREF.LOAN_DEFAULT	Observations	1000
Member Type	DATA	Variables	16
Engine	V9	Indexes	0
Created	10/04/2017 13:52:55	Observation Length	128
Last Modified	10/04/2017 13:52:55	Deleted Observations	0
Protection		Compressed	NO
Data Set Type		Sorted	NO
Label			
Data Representation	SOLARIS_X86_64, LINUX_X86_64, ALPHA_TRU64, LINUX_IA64		
Encoding	utf-8 Unicode (UTF-8)		

Alphabetic List of Variables and Attributes

#	Variable	Type	Len	Format	Informat
15	Age	Num	8	BEST12.	BEST32.
12	Amount	Num	8	BEST12.	BEST32.
7	Car_loan	Num	8	BEST12.	BEST32.
2	Checking_amount	Num	8	BEST12.	BEST32.
4	Credit_score	Num	8	BEST12.	BEST32.
1	Default	Num	8	BEST12.	BEST32.
10	Education_loan	Num	8	BEST12.	BEST32.
14	Emp_duration	Num	8	BEST12.	BEST32.
11	Emp_status	Char	10	$10.	$10.
5	Gender	Char	6	$6.	$6.
9	Home_loan	Num	8	BEST12.	BEST32.
6	Marital_status	Char	7	$7.	$7.

(continued)

Alphabetic List of Variables and Attributes

#	Variable	Type	Len	Format	Informat
16	No_of_credit_acc	Num	8	BEST12.	BEST32.
8	Personal_loan	Num	8	BEST12.	BEST32.
13	Saving_amount	Num	8	BEST12.	BEST32.
3	Term	Num	8	BEST12.	BEST32

```
/*Descriptive statistics of the data*/
proc means data = libref.loan_default;
var Term Saving_amount ;
run;
```

Applying proc means that the descriptive statistic or summary of the data is displayed like the Number of observations (N), Mean, Standard deviation; and the Min and Max values of the respective variable are represented.

Table 2-5.

The MEANS Procedure

Variable	N	Mean	Std Dev	Minimum	Maximum
Term	1000	17.8150000	3.2405673	9.0000000	27.0000000
Saving_amount	1000	3179.27	339.5497508	2082.00	4108.00

```
/* Applying Proc freq to see the frequency of the data */

proc freq data = libref.loan_default ;
tables Emp_status Default Default * Emp_status;
run;
```

Procedure frequency represents the number of frequency of each levels with cumulative frequency and cumulative percentage, like the number of employed, number of unemployed, number of non-default (0), and number of default (1).

Table 2-6.

The FREQ Procedure

Emp_status	Frequency	Percent	Cumulative Frequency	Cumulative Percent
Employed	308	30.80	308	30.80
Unemployed	692	69.20	1000	100.00

Default	Frequency	Percent	Cumulative Frequency	Cumulative Percent
0	700	70.00	700	70.00
1	300	30.00	1000	100.00

Interaction between two variables like default and Emp_status display the table of default by Emp_status with frequency, percent, row percent, and column percent details.

Frequency Percent Row Pct Col Pct	Table of Default by Emp_status		
	Emp_status		
Default	employed	unemployed	Total
0	218	482	700
	21.80	48.20	70.00
	31.14	68.86	
	70.78	69.65	
1	90	210	300
	9.00	21.00	30.00
	30.00	70.00	
	29.22	30.35	
Total	308	692	1000
	30.8	69.20	100.00

```
/* Applying proc univariate to get more detailed summary of the data */
proc univariate data = libref.loan_default;
var Saving_amount;
histogram  Saving_amount/normal;
run;
```

In this case, partial output of procedure univariate is displayed in Table 2-7.

Table 2-7.

Part 1

The UNIVARIATE Procedure
Variable: Saving_amount
Moments

N	1000	Sum Weights	1000
Mean	3179.266	Sum Observations	3179266
Std Deviation	339.549751	Variance	115294.033
Skewness	-0.1555899	Kurtosis	-0.1041112
Uncorrected SS	1.02229E10	Corrected SS	115178739
Coeff Variation	10.6801303	Std Error Mean	10.7375059

Proc univariate is one of the procedures that is used to display the detailed summary or descriptive statistic of the data. Part 1 will display kurtosis, skewness, standard deviation, uncorrected ss, corrected ss, std error mean, etc.

Part 2 will display more basic statistics like mean, median, mode, std deviation, variance, range. and interquartile range.

Part 2

Basic Statistical Measures

Location		Variability	
Mean	3179.266	Std Deviation	339.54975
Median	3203.000	Variance	115294
Mode	3201.000	Range	2026
		Interquartile Range	451.50000

Note: The mode displayed is the smallest of 5 modes with a count of 5.

Part 3 will display the Test column that lists the various tests like student's t –test, sign test, and signed rank test. The second column is the statistic column, which lists the values of the test statistics; and the third column is the p value column, which lists the p values associated with the test statistics. In this case, all the tests values and the corresponding p-values are less than 0.0001, which conclude that the variable is statistically significant.

Part 3

Tests for Location: Mu0=0				
Test		Statistic	p Value	
Student's t	T	296.0898	Pr > ltl	<.0001
Sign	M	500	Pr >= lMl	<.0001
Signed Rank	S	250250	Pr >= lSl	<.0001

Part 4 will display the quantiles like 100% Max, 99%, 95%, 90%, 75%Q3, etc.

Part 4

Quantiles (Definition 5)	
Level	Quantile
100% Max	4108.0
99%	3971.5
95%	3716.5
90%	3606.5
75% Q3	3402.5
50% Median	3203.0
25% Q1	2951.0
10%	2728.5
5%	2613.5
1%	2356.0
0% Min	2082.0

Part 5 will display the five lowest and five highest values of the variable.

Part 5

Extreme Observations

Lowest		Highest	
Value	**Obs**	**Value**	**Obs**
2082	616	4014	146
2145	740	4021	608
2191	532	4022	659
2191	14	4044	407
2248	971	4108	373

It is also used to assess the normal distribution of the data. Figure 2-7 displays the normal distribution of the saving amount present in the data.

The UNIVARIATE Procedure

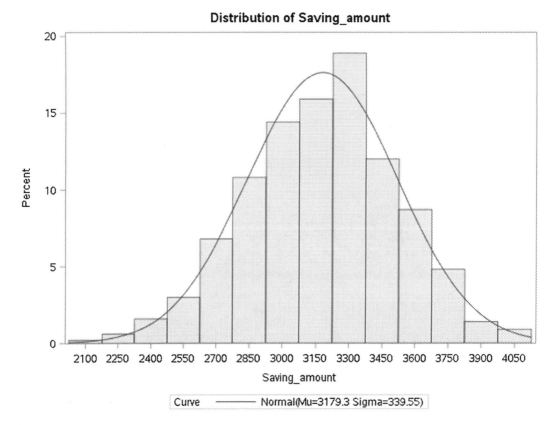

Figure 2-7. *Normal distribution for Saving_amount*

Part 6 will display the histogram for the variable Saving_amount with a normal curve as shown in Figure 2-7. The estimated parameter for the normal distributions curve lists in parameter, symbol, and estimate column. In the parameter column, Mean = 3179.27 and Std Dev = 339.55 is displayed for the Saving_amount variable.

Part 6

The UNIVARIATE Procedure		
Fitted Normal Distribution for Saving_amount		
Parameters for Normal Distribution		
Parameter	Symbol	Estimate
Mean	Mu	3179.266
Std Dev	Sigma	339.5498

Part 7 displays the three goodness-of-fit tests for Normal Distribution: the Kolmogorov-Smirnov, the Cramér–von Mises, and the Anderson-Darling tests. p values of all the tests are less than 0.05, which concludes that the fitted normal distribution is a good model for the distribution of Saving_amount.

Part 7

Goodness-of-Fit Tests for Normal Distribution				
Test	Statistic		p Value	
Kolmogorov-Smirnov	D	0.03447951	Pr > D	<0.010
Cramér-von Mises	W-Sq	0.14736169	Pr > W-Sq	0.026
Anderson-Darling	A-Sq	0.77278213	Pr > A-Sq	0.046

```
/* Applying proc corr to find out the correlation between Variables in the data */
proc corr data = libref.loan_default;
var Default Age;
run;
```

Pearson Correlation Coefficients display the correlation between the variables so here the correlation between Default and Age is (**-0.66445**) Negative correlation, which displays the bank-loan default, will decrease with the increase in Age as both variables are indirectly proportional to each other. The correlation between other variables can be calculated using the same approach.

Table 2-8.

The CORR Procedure

2 Variables: Default Age

Simple Statistics

Variable	N	Mean	Std Dev	Sum	Minimum	Maximum
Default	1000	0.30000	0.45849	300.00000	0	1.00000
Age	1000	31.20900	4.09317	31209	18.00000	42.00000

Pearson Correlation Coefficients, N = 1000 Prob > |r| under H0: Rho=0

	Default	**Age**
Default	1.00000	-0.66445
		<.0001
Age	-0.66445	1.00000
	<.0001	

Model Building and Interpretation of Full Data

In this section, we are using SAS Studio and complete data is used for model building and interpretation; and in the next section, we are randomly splitting data into two parts: training data set and testing data set. Training data set is used to build the model and testing data set is used to test the performance of the model. There are variations seen in the SAS output of both the model (Full and Training) due to the sampling.

In program1 code the **Descending** option is mentioned so SAS would model **Default =1**. If the descending option is omitted, then SAS would model **Default = 0** and the result of the complete model would be reversed. Instead of getting the probability for **Default =1**, the model would have predicted the probability of **Default = 0**. Therefore in **proc logistic** it is essential to add the **Descending** option because by default **proc logistic** model 0s rather than 1s. **Class** statement is added when there are categorical variables in the data, so in order to include categorical variables in our model, the Class statement is used The **param = effect ref = first** option after the slash is used so it can match with the R result where by default the first category of the categorical variable is omitted. Default is our **dependent variable** in the data and Checking_amount, Term Credit_score, Car_loan, Personal_loan, Home_loan, Education_loan, Amount, Saving_ amount, Emp_duration, Gender, Marital_status, Age, No_of_credit_acc, and Emp_status are all **independent variables** in our model. Link = logit is used by default in logistic regression as it is binomial family. **Score** statement is used so the fitted model can be used to score the **loan_default** data, which is present in our working library libref. The final output is generated where we can see that the data together with the predicted values are displayed and are saved in the **dataset Logistic_result.**

```
/*Building logistic regression model on full data */
```
Program1:

```
proc logistic data = libref.loan_default descending ;
class Gender Marital_status Emp_status
/ param=effect ref=first;
model default = Checking_amount Term Credit_score Car_loan Personal_loan
Home_loan Education_loan Amount Saving_amount Emp_duration Gender
Marital_status Age No_of_credit_acc Emp_status        / link=logit;
score out = Logistic_result ;
run;
```

Partial output is shown in Table 2-9, which displays the coefficients (labeled Estimate), their standard errors (error), the Wald Chi-Square statistic, and associated p-values. Table 2-9 displays that the variables where p values are less than 0.05, only those variables are playing the significant role in the model so removing all other insignificant variables where p values are more than 0.05. The model is rebuilt only using significant variables like **Checking_amount, Term, Credit_score, Saving_ amount,** and **Age**.

Table 2-9. *Coeffecient Table*

Analysis of Maximum Likelihood Estimates

Parameter		DF	Estimate	Standard Error	Wald Chi-Square	Pr > ChiSq
Intercept		1	40.2003	4.7469	71.7210	<.0001
Checking_amount		1	-0.00509	0.000676	56.6661	<.0001
Term		1	0.1704	0.0521	10.7029	0.0011
Credit_score		1	-0.0110	0.00207	28.0043	<.0001
Car_loan		1	-0.6005	2.7588	0.0474	0.8277
Personal_loan		1	-1.5541	2.7588	0.3173	0.5732
Home_loan		1	-3.5684	2.8460	1.5721	0.2099
Education_loan		1	0.6499	2.7898	0.0543	0.8158
Amount		1	0.000803	0.000511	2.4630	0.1166
Saving_amount		1	-0.00482	0.000609	62.7550	<.0001
Emp_duration		1	0.00292	0.00444	0.4320	0.5110
Gender	Male	1	0.0975	0.2548	0.1465	0.7019
Marital_status	Single	1	0.1676	0.2460	0.4639	0.4958
Age		1	-0.6475	0.0647	100.2633	<.0001
No_of_credit_acc		1	-0.0969	0.1007	0.9261	0.3359
Emp_status	Unemployed	1	0.2936	0.1737	2.8566	0.0910

In program1.1 code Default is **dependent variable** in the data and Checking_ amount, Term, Credit_score, Saving_amount, and Age are the previously identified significant **independent variables** in our model. **Score** statement is used so the fitted model can be used to score the **loan_default** data, which is present in our working library libref. The final output is generated where we can see that the data together with the predicted values are displayed and are saved in the **dataset Logistic_result**.

/*Rebuilding logistic model on full data after removing insignificant variables from the data */

/*Program1.1*/

```
proc logistic data = libref.loan_default descending;
model default = Checking_amount Term Credit_score
Saving_amount Age/ link=logit;
score out = Logistic_result ;
run;
```

The **Logistic regression output of program1.1** is split into several parts, each of which is discussed in the following sections.

The first and second part of the program1.1 output displays the loan_default file that is being analyzed and the total number of observations read and used in our analysis is equal to 1000. Default is the dependent or target variable in our analysis, SAS is the modeling Default using the binary logit model, and the optimization technique is Fisher's scoring.

Part 1

The LOGISTIC Procedure	
Model Information	
Data Set	LIBREF.LOAN_DEFAULT
Response Variable	Default
Number of Response Levels	2
Model	binary logit
Optimization Technique	Fisher's scoring

Part 2

Number of Observations Read	1000
Number of Observations Used	1000

Part 3 of the program1.1 display that the probability of Default = 1 is being modeled that is displayed by using the descending option in our code. If the descending option is omitted then SAS would model Default = 0 and the result of the complete model would be reversed. Instead of getting the probability for default =1, the model would have predicted the probability of Default = 0. Therefore, in proc logistic it is essential to add the descending option because by default proc logistic models 0s rather than 1s.

Part 3

Response Profile		
Ordered Value	Default	Total Frequency
1	1	300
2	0	700

Probability modeled is Default='1'.

Model Convergence Status

Convergence criterion (GCONV=1E-8) satisfied.

Part 4 of the program1.1 Model Fit Statistics section displays the model fit output that will help in assessing the overall fit of the model . The Full model -2 log L (1221.729) would be used for the comparison with the Nested model. The AIC and SC are the information criteria used in the model.

Part 4

Model Fit Statistics		
Criterion	Intercept Only	Intercept and Covariates
AIC	1223.729	351.970
SC	1228.636	381.416
-2 Log L	1221.729	339.970

The next section, Part 5 of program1.1 displays about the likelihood ratio, score, and Walt test . The likelihood ratio Chi-Square of **(881.7588)** with 5 degress of freedom and associated P value of **0.0001** signify that our full model fits significantly better than our Null model . Similarly the score test Chi-Square of **(635.5404)** with 5 degress of freedom and associated P value of **0.0001** and Wald test Chi-Square of **(186.1945)** with 5 degress of freedom and associated P value of **0.0001** also signify that the model is statistically significant.

Part 5

Testing Global Null Hypothesis: BETA=0			
Test	Chi-Square	DF	Pr > ChiSq
Likelihood Ratio	881.7588	5	<.0001
Score	635.5404	5	<.0001
Wald	186.1945	5	<.0001

Table 2-10 (part 6 from program1.1) displays the coefficients (labeled Estimate), their standard errors (error), the Wald Chi-Square statistic, and associated p-values. Coefficients Table 2-10 displays all the significant variables where p values are less than 0.05 like **Checking_amount, Term, Credit_score, Saving_amount and Age**.

The coefficients for Checking_amount, Term, Credit_score, Saving_amount, and Age are statistically significant as their P values are less than 0.05. In the logistic regression model, for a one-unit increase in the predictor variables, it will display the change (increase or decrease) in the log odds of the outcome. It can be explained as:

1. For every one-unit change in **Checking_amount**, the log odds of bank-loan default versus bank- loan non-default decreases by **(-0.004)**.

2. Similarly it is for **Term** that one-unit change in Term, the log odds of bank-loan default versus bank-loan non-default increases by *(0.174)*.

3. For **Credit_score** one-unit change in Credit_score, the log odds of bank-loan default versus bank-loan non-default decreases by **(-0.011)**.

4. For **Saving_amount** one-unit change in Saving_amount, the log odds of bank-loan default versus bank-loan non-default decreases by **(-0.004)**.

5. For **Age** one-unit change in Age, the log odds of bank-loan default versus bank-loan non-default decreases by **(-0.628)**.

Note   The Coefficients table values for the binomial logistic model in both R and SAS tools would always be similar when the model is built on the full data.

Table 2-10. *Coeffecient Table*

Part 6

Analysis of Maximum Likelihood Estimates

Parameter	DF	Estimate	Standard Error	Wald Chi-Square	Pr > ChiSq
Intercept	1	38.8469	3.5109	122.4281	<.0001
Checking_amount	1	-0.00484	0.000618	61.3643	<.0001
Term	1	0.1748	0.0474	13.6281	0.0002
Credit_score	1	-0.0114	0.00198	33.2773	<.0001
Saving_amount	1	-0.00451	0.000552	66.9354	<.0001
Age	1	-0.6286	0.0588	114.4127	<.0001

Part 7 of program1.1 displays the odds ratio estimate for each significant variable and the corresponding 95% Wald confidence intervals for variables without interactions An odd ratio is the exponentiated coefficient like, for example, the **Checking_amount** coefficient value is **-0.00484** then **exp(0.00484) is 0.995**. Similarly for **Term** coefficient value is **0.1748** then **exp(0.1748) is 1.191** and similarly other variable exponentiated coefficient is calculated.

It can be interpreted as, for example, for a one-unit increase in **Checking_amount**, the odds of being bank-loan Default (versus bank-loan non-default) decreases by a factor of **0.995** and similarly for a one-unit increase in **Term**, the odds of being bank-loan Default (versus bank-loan non-default) increases by a factor of **1.191**.

Part 7

Odds Ratio Estimates

Effect	Point Estimate	95% Wald Confidence Limits	
Checking_amount	0.995	0.994	0.996
Term	1.191	1.085	1.307
Credit_score	0.989	0.985	0.993
Saving_amount	0.995	0.994	0.997
Age	0.533	0.475	0.598

Part 8 of program1.1 displays the association of predicted probabilities and observed responses. The above section displays several measures for assessing the predictive accuracy of the model for how well the model is predicting the target or dependent variable. Concordant is defined as when the observation of a non-default bank-loan (default=0) has a lower predicted probability than the observation with bank-loan default (default=1). On the other side Discordant is defined as when the observation of a non-default bank-loan (default=0) has a higher predicted probability than the observation with bank-loan default (default=1) and the observations where the predicted probability for non-default bank loan and default bank loan is same is considered as Tied.

Model is considered as the better model in term of predictive power when there is a high percentage of Concordant pairs and low percentage of Discordant pairs. Pairs are defined as all possible pairs of observations as in this data set of loan default; there are non-default loan application equal to 700 and default loan applications equal to 300. Thus **700*300 = 210000**. In proc logistic, there are four rank correlation indexes like **Somers'D, Gamma, Tau-a, and C statistic**. These are the tests that are used for assessing the predictive ability of a model. Among these four tests, C statistic is commonly used to assess the predictive accuracy of the model, and the higher the C statistic value, the higher is the predictive accuracy of the model.

Part 8

Association of Predicted Probabilities and Observed Responses

Percent Concordant	97.8	Somers' D	0.956
Percent Discordant	2.2	Gamma	0.956
Percent Tied	0.0	Tau-a	0.402
Pairs	210000	C	0.978

```
/*splitting datasets into training (70%) and testing (30%) */

proc surveyselect data= libref.loan_default
method=srs seed=2 outall
samprate=0.7  out=libref.credit_subset;

proc print data=libref.credit_subset;
  run;
```

In the Survey, select Procedure section; we split data by using the Simple random sampling method and part 9 displays sampling rate as 0.7 and seed is set at 2 (same seed value used in R). We are setting seed value to reproduce the same sample and Credit_ subset is the output dataset after sampling is conducted. We can run proc print to print the Credit_subset output file.

The SURVEYSELECT Procedure

Selection Method	Simple Random Sampling

Part 9

Input Data Set	LOAN_DEFAULT
Random Number Seed	2
Sampling Rate	0.7
Sample Size	700
Selection Probability	0.7
Sampling Weight	0
Output Data Set	CREDIT_SUBSET

Now divide the data into training and testing data set based on the selection variable. Where selection is equal to 1, assign all those observations to training data set and where selection is equal to 0, assign all those observations to testing data set.

```
/* Values of selected variable 1 means selected for training, 0 means
selected for testing data set */

data libref.training;
set libref.credit_subset;
if selected=1;
proc print;

data libref.testing;
set libref.credit_subset;
if selected=0;
proc print;
```

```
/*apply proc freq in training and testing data to Check for  balanced data
for target var */

proc freq data = libref.training;
tables Default;
run;
```

Procedure frequency table displays that in training data set out of 700 observations, there are 490 observations for Default = 0 and 210 observations for Default = 1.

The FREQ Procedure

Default	Frequency	Percent	Cumulative Frequency	Cumulative Percent
0	490	70.00	490	70.00
1	210	30.00	700	100.00

```
proc freq data = libref.testing;
tables Default;
run;
```

Procedure frequency table displays that in testing data set out of 300 observations, there are 210 observations for Default = 0 and 90 observations for Default = 1.

The FREQ Procedure

Default	Frequency	Percent	Cumulative Frequency	Cumulative Percent
0	210	70.00	210	70.00
1	90	30.00	300	100.00

In Program2 code **Descending** option, **Class** statement, **param = effect ref = first are, Link = logit** are used for reasons explained previously for Program1. Default is our **dependent variable** in the data and Checking_amount, Term Credit_score, Car_loan, Personal_loan, Home_loan, Education_loan, Amount, Saving_amount, Emp_duration, Gender, Marital_status, Age, No_of_credit_acc and Emp_status are all **independent variables** in the current model. **Score** statement is used so the fitted model can be used to score the **testing data** that is present in our working library libref. The final output is generated where we can see that the data together with the predicted values are displayed and are saved in the dataset **Logistic_output**.

/*Building logistic regression model on training data set */

/*Program2*/

```
ODS GRAPHICS ON;
proc logistic data = libref.training
descending PLOTS (ONLY) = ROC ;
class Gender Marital_status Emp_status
/param=effect ref=first;
model default = Checking_amount Term Credit_score Car_loan Personal_
loan      Home_loan      Education_loan Amount      Saving_amount Emp_
duration Gender      Marital_status      Age No_of_credit_acc Emp_
status      / link=logit;
score data = libref.testing out=WORK.Logistic_output;
run;
ODS GRAPHICS OFF;
```

Table 2-11 displays the coefficients (labeled Estimate), their standard errors (error), the Wald Chi-Square statistic, and associated p-values. Coefficients Table 2-5 displays that the variables where p values are less than 0.05 only those variables are playing the significant role in the model so removing all other insignificant variables where p values are more than 0.05. The model is rebuilt using significant variables **Checking_amount, Term, Credit_score, Amount, Saving_amount, Age and Emp_status.**

Table 2-11. *Coeffecient Table*

Analysis of Maximum Likelihood Estimates

Parameter	DF	Estimate	Standard Error	Wald Chi-Square	Pr > ChiSq
Intercept	1	38.8716	6.2033	39.2658	<.0001
Checking_amount	1	-0.00487	0.000794	37.7006	<.0001
Term	1	0.1766	0.0643	7.5517	0.0060
Credit_score	1	-0.0122	0.00258	22.4641	<.0001
Car_loan	1	-1.1416	4.2380	0.0726	0.7876
Personal_loan	1	-2.3299	4.2369	0.3024	0.5824

(continued)

Table 2-11. (*continued*)

Analysis of Maximum Likelihood Estimates

Parameter		DF	Estimate	Standard Error	Wald Chi-Square	Pr > ChiSq
Home_loan		1	-4.6337	4.4005	1.1088	0.2923
Education_loan		1	-0.2952	4.2672	0.0048	0.9448
Amount		1	0.00140	0.000641	4.7757	0.0289
Saving_amount		1	-0.00444	0.000717	38.3498	<.0001
Emp_duration		1	-0.00091	0.00545	0.0281	0.8668
Gender	Male	1	0.0382	0.6184	0.0038	0.9507
Marital_status	Single	1	0.3143	0.5850	0.2887	0.5911
Age		1	-0.6318	0.0761	68.9736	<.0001
No_of_credit_acc		1	-0.1440	0.1196	1.4489	0.2287
Emp_status	Unemployed	1	1.0806	0.4364	6.1327	0.0133

```
/* Building logistic model with significant variables */
/*Program2.1*/

ODS GRAPHICS ON;
proc logistic data = libref.training
descending PLOTS (ONLY) = ROC ;
class Emp_status/ param=ref ref=first;
model default = Checking_amount Term Credit_score Emp_status
Amount Saving_amount Age/ link=logit  ;
score data = libref.testing out=WORK.Logistic_output;
run;
ODS GRAPHICS OFF;
```

Table 2-12 displays the coefficients (labeled Estimate), their standard errors (error), the Wald Chi-Square statistic, and associated p-values. Table 2-12 displays the variables where p values are less than 0.05 and only those variables are playing the significant role in the model so removing Amount whose p values is more than 0.05. The final model is rebuilt using significant variables **Checking_amount, Term, Credit_score, Saving_amount, Age, and Emp_status**.

Table 2-12. *Coeffecient Table*

Analysis of Maximum Likelihood Estimates

Parameter		DF	Estimate	Standard Error	Wald Chi-Square	Pr > ChiSq
Intercept		1	36.0641	4.0925	77.6561	<.0001
Checking_amount		1	-0.00477	0.000738	41.8069	<.0001
Term		1	0.1919	0.0571	11.2911	0.0008
Credit_score		1	-0.0121	0.00249	23.4346	<.0001
Emp_status	Unemployed	1	0.8525	0.3811	5.0030	0.0253
Amount		1	0.000846	0.000598	2.0008	0.1572
Saving_amount		1	-0.00410	0.000650	39.7314	<.0001
Age		1	-0.6251	0.0713	76.8687	<.0001

```
/* Rebuilding Final Logistic Model with only significant variables */
/*Program2.2*/

ODS GRAPHICS ON;
proc logistic data = libref.training
descending PLOTS (ONLY) = ROC ;
class Emp_status/ param=ref ref=first;
model default = Checking_amount Term Credit_score Emp_status
Saving_amount Age/ link=logit  ;
score data = libref.testing out=WORK.Logistic_output;
run;
ODS GRAPHICS OFF;
```

The **Logistic regression output of Program2.2** is split into several sections, and each is discussed below

The first and second part of program2.2 output display the training data file that is being analyzed, and the total number of observations read and used in our analysis is equal to 700. Default is the dependent or target variable in our analysis and SAS is modeling Default using the binary logit model and the optimization technique is Fisher's scoring.

Part 1

The LOGISTIC Procedure
Model Information

Data Set	LIBREF.TRAINING
Response Variable	Default
Number of Response Levels	2
Model	binary logit
Optimization Technique	Fisher's scoring

Part 2

Number of Observations Read	700
Number of Observations Used	700

Part 3 of program2.2 displays the probability of Default = 1 being modeled, which is displayed by using the descending option in our code

Part 3

Response Profile

Ordered Value	Default	Total Frequency
1	1	210
2	0	490

Probability modeled is Default='1'.

Part 4 of program2.2 display thats in Emp_status the Employed is represented as 0 and unemployed as 1.

Part 4

Class Level Information

Class	Value	Design Variables
Emp_status	**Employed**	0
	Unemployed	1

Model Convergence Status

Convergence criterion (GCONV=1E-8) satisfied.

The Model Fit Statistics part 5 section from program2.2 displays the model fit output, which will help in assessing the overall fit of the model . The Full model -2 log L **(885.210)** would be used for the comparison with the Nested model . The AIC and SC are the information criteria used in the model.

Part 5

Model Fit Statistics

Criterion	Intercept Only	Intercept and Covariates
AIC	857.210	254.069
SC	861.761	285.926
-2 Log L	855.210	240.069

Part 6 of the section display is about the likelihood ratio, score, and Walt test. The likelihood ratio Chi-Square of **(615.1415)** with 6 degress of freedom and associated P value of **0.0001** signify that our full model fits significantly better than our Null model. Similarly the score test Chi-Square of **(443.5417)** with 6 degress of freedom and associated P value of **0.0001** and Wald test Chi-Square of **(133.8054)** with 6 degress of freedom and associated P value of **0.0001** also signify that the model is statistically significant.

Part 6

Testing Global Null Hypothesis: BETA=0

Test	Chi-Square	DF	Pr > ChiSq
Likelihood Ratio	615.1415	6	<.0001
Score	443.5417	6	<.0001
Wald	133.8054	6	<.0001

Type 3 Analysis of Effects part 7 section from program2.2 displays the test for each of the variables in the model individually. The Chi-Square test statistics and associated p-values (p- values less than 0.05) signify that each of the six variables in the model significantly improve the model fit.

Part 7

Type 3 Analysis of Effects

Effect	DF	Wald Chi-Square	Pr > ChiSq
Checking_amount	1	42.3971	<.0001
Term	1	11.4042	0.0007
Credit_score	1	22.7258	<.0001
Emp_status	1	5.2419	0.0220
Saving_amount	1	39.5409	<.0001
Age	1	79.5906	<.0001

Part 8 of the program2.2 represents Coefficients Table 2-13, which displays the coefficients (labeled Estimate), their standard errors (error), the Wald Chi-Square statistic, and associated p-values. Table 2-13 displays all the significant variables where p values are less than 0.05 like **Checking_amount, Term, Credit_score, Emp_status Saving_amount, and Age**. The coefficients for Checking_amount, Term, Credit_score, Emp_status Saving_amount, and Age are statistically significant as their P values are less than 0.05. In the logistic regression model for a one-unit increase in the predictor variables, it will display the change (increase or decrease) in the log odds of the outcome. It can be explained as:

1. For every one-unit change in **Checking_amount**, the log odds of bank-loan default versus bank-loan non-default decreases by **(-0.004)**.

2. Similarly it is for **Term** that one-unit change in Term, the log odds of bank-loan default versus bank-loan non-default increases by *(0.174)*.

3. The coefficients for the categories of **Emp_status** would have different interpretations. For example, having **Emp_status** with **unemployed**, versus with **employed**, increases the log odds of bank-loan default by **(0.871).**

Table 2-13. *Coefficient Table*

Part 8

Analysis of Maximum Likelihood Estimates

Parameter		DF	Estimate	Standard Error	Wald Chi-Square	Pr > ChiSq
Intercept		1	37.0413	4.0704	82.8109	<.0001
Checking_amount		1	-0.00476	0.000731	42.3971	<.0001
Term		1	0.1919	0.0568	11.4042	0.0007
Credit_score		1	-0.0119	0.00249	22.7258	<.0001
Emp_status	Unemployed	1	0.8711	0.3805	5.2419	0.0220
Saving_amount		1	-0.00409	0.000651	39.5409	<.0001
Age		1	-0.6296	0.0706	79.5906	<.0001

Part 9 section of odd ratio estimate displays the point estimate for each significant variables and the corresponding 95% Wald confidence intervals for variables without interactions An odd ratio is the exponentiated coefficient like, for example, the **Checking_amount** coefficient value is **-0.00476** then **exp(0.00476) is 0.995**. Similarly for **Term** the coefficient value is **0.1919 and** then **exp(0.1919) is 1.212** and similarly another variable exponentiated coefficient is calculated.

It can be interpreted as, for example, for a one-unit increase in **Checking_amount**, the odds of being bank-loan Default (versus bank-loan non-default) decreases by a factor of **0.995** and similarly for a one-unit increase in **Term** the odds of being bank-loan Default (versus bank-loan non-default) increases by a factor of **1.212**.

Part 9

Odds Ratio Estimates

Effect	Point Estimate	95% Wald Confidence Limits	
Checking_amount	0.995	0.994	0.997
Term	1.212	1.084	1.354
Credit_score	0.988	0.983	0.993
Emp_status unemployed vs employed	2.389	1.134	5.037
Saving_amount	0.996	0.995	0.997
Age	0.533	0.464	0.612

The part 10 section of Association of predicted probabilities and observed response displays several measures for assessing the predictive accuracy of the model and how well the model is predicting the target or dependent variable.

Part 10

Association of Predicted Probabilities and Observed Responses

Percent Concordant	97.7	Somers' D	0.953
Percent Discordant	2.3	Gamma	0.953
Percent Tied	0.0	Tau-a	0.401
Pairs	102900	C	0.977

Receiver Operating Characteristic (ROC) and Area Under Curve is displayed in Figure 2-8. On the y-axis (sensitivity) is displayed and x-axis (1- specificity) is displayed, the curve above the diagonal represents the better performance; below the diagonal, it represents the worse performance and approaching toward 1 it represents the best performance.

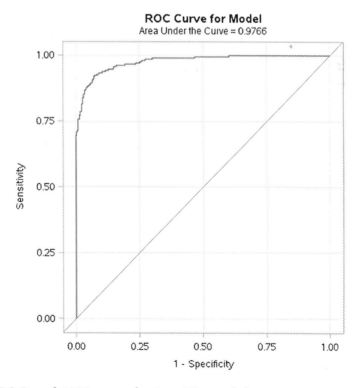

Figure 2-8. ROC and AUC curve for Logsitic model

The Area Under Curve (AUC) values varies from range 0.5 to 1.0. Where 0.5 means the worse or poor predictive ability and 1.0 is considered as the accurate or best predictive ability. The graph approaching toward 1 signifies that the model predictive accuracy is high. The higher the AUC value, the better he model predictive accuracy; and the lower the AUC value, the model predictive accuracy is worse. In this model AUC value is **97%,** which signifies high predictive accuracy of the model.

Summary

In this chapter we learned about application of data analytics in the banking sector, which is one of the first adopters of computers and data analytics. The model discussed in the chapter is the logistic regression model. Various characteristics, features, and assumptions of the model were discussed. Practical application of this model in a real-life scenario was demonstrated by help of a case study to predict bank-loan defaults, which is a big challenge for the banking sector, and it can help the bank better understand its customers in terms of lending bank loans. Model development, execution, visualization, result interpretations, and evaluation was also performed by using both R and SAS Studio.

References

1. Morisi, T. L. Commercial Banking transformed by Computer technology *Monthly Labor Review* [Online], 1996, pp. 30–36.

2. Fisher, E. W.; McKenney, J. L., The development of ERMA Banking system: Lessons from History. *IEEE Annals of the History of Computing* **1993,** *15* (1), pp. 44–57.

3. International convergence of capital measurements and capital standards. Supervision, B. C. o. B., Ed. Bank of International Settlements: Basel, Switzerland, 1988.

4. International Convergence of Capital Measurement and Capital Standards A Revised Framework Comprehensive Version. Supervision, B. C. o. B., Ed. Bank of international settlements: http://www.bis.org/publ/bcbs128.pdf, 2006.

5. Basel III: A global regulatory framework for more resilient banks and banking systems. Supervision, B. C. o. b., Ed. Bank of International Settlements: `http://www.bis.org/publ/bcbs189.pdf`, 2010 (Rev 2011).

6. Ginovsky, J. Analytical eye brings greater customer insight. `http://www.bankingexchange.com/blogs-3/reporter-s-notebook/item/6826-analytical-eye-brings-greater-customer-insight` (accessed September 17, 2017).

7. SAS Banking Analytics Customer stories. (accessed October 9, 2017).

8. Wibke, H. Upselling or Upsetting? Studies on the Behavioral Consequences of Upsell Offers in Service Encounters. The University of St. Gallen, Germany, 2012.

9. Toos, D., How advanced analytics are redefining banking. McKinsey& Company: `http://www.mckinsey.com/business-functions/digital-mckinsey/our-insights/how-advanced-analytics-are-redefining-banking`, 2013; p 3.

10. Marous, J., 9 Keys to Bank Cross-Selling Success. FinExtra: `https://www.finextra.com/blogposting/7829/9-keys-to-bank-cross-selling-success`, 2013; Vol. 2017.

11. Woods, R. Z. J. A., *Best Practices in Customer Service*. AMACOM Div American Mgmt Assn.

12. Neslin, S. A.; Gupta, S.; Kamakura, W.; Lu, J.; Mason, C. H., Defection Detection: Measuring and Understanding the Predictive Accuracy of Customer Churn Models. *Journal of Marketing Research* **2006**, *43* (2), 204–211.

13. He, B.; Shi, Y.; Wan, Q.; Zhao, X., Prediction of Customer Attrition of Commercial Banks based on SVM Model. *Procedia Computer Science* **2014**, *31* (Supplement C), 423–430.

14. Danielle, D.; V., D. J., The Rise and Fall of Subprime Mortgages. *Economic Letter* 2007, pp 1–8.

15. Darryl, G. E., U.S. Implementation of the Basel Capital Regulatory Framework. Service, C. R., Ed. 2014; p 22.

16. *Banking Systems.* 2 ed.; Cengage Learning: USA, 2009; p 448.

17. Anuj, S.; Prabin, P. K., A review of financial accounting fraud detection based on data mining techniques. *International Journal of computer applications* **2013**, *39* (1), 37–47.

18. Madinger, J., *Money Laundering: A Guide for Criminal Investigators, Third Edition.* 3 illustrated ed.; CRC Press: USA, 2016; p. 430.

19. Nash, J. R., *The Great Pictorial History of World Crime.* Scarecrow Press: 2004; Vol. 2, p. 1200.

20. Tim, W., The great Moldovan bank robbery. *BBC News,* June 18, 2015.

21. Kottasova, I., Global banks handled laundered Russian cash worth hundreds of millions. *CNN Money* March 24, 2017.

22. Young, R. D.; Rice, T. How do banks make money? The fallacies of fee income *Economic prospectives* [Online], 2004, p. 18.

23. Hilbe, J. M., *Practical Guide to Logistic Regression.* CRC Press: 2016, p 174.

24. Michael, B. A., What You See May Not Be What You Get: A Brief, Nontechnical Introduction to Overfitting in Regression-Type Models. *Psychosomatic Medicine* **2004**, *66*, 411–421.

25. Farrar, D. E.; Glauber, R. R., Multicollinearity in Regression Analysis: The Problem Revisited. *The Review of Economics and Statistics* **1967**, *49* (1), 92-107.

26. Briggs, N. E.; MacCallum, R. C., Recovery of Weak Common Factors by Maximum Likelihood and Ordinary Least Squares Estimation. *Multivariate Behavioral Research* **2003**, *38* (1), 25–56.

27. Bewick, V.; Cheek, L.; Ball, J., Statistics review 14: Logistic regression. *Critical Care* **2005**, *9* (1), 112.

28. Menard, S., *Applied Logistic Regression Analysis.* SAGE Publications: 2002.

29. Krzanowski, W. J.; Hand, D. J., *ROC Curves for Continuous Data.* CRC Press: 2009.

30. Sim, J.; Lee, J. S.; Kwon, O., Missing Values and Optimal Selection of an Imputation Method and Classification Algorithm to Improve the Accuracy of Ubiquitous Computing Applications. *Mathematical Problems in Engineering* **2015**, *2015*, 14.

31. Sedgwick, P., Pearson's correlation coefficient. *BMJ: British Medical Journal* **2012**, *345*.

32. Midi, H.; Sarkar, S. K.; Rana, S., Collinearity diagnostics of binary logistic regression model. *Journal of Interdisciplinary Mathematics* **2010**, *13* (3), 253–267.

## CHAPTER 3

Retail Case Study

Retailing is defined as the act of purchasing the products and services from manufacturers and wholesalers in large quantities and selling and renting those products and services to the individual consumer for their personal and household use.[1] In the retailing industry, the goods and services can be sold to the consumers by various channels like in-store sales, online sales, through the mail, over the phone, and door-to-door sales. The retail sector is one of the largest private sector industries in the world and plays a vital role in the economy's success. Figures from the US Census Bureau indicate that in the United States alone, retail trade revenue was over 4.7 trillion USD[2] in 2015.

Traditional retailing was about the small and local stores, but today's retailing has been completely transformed by globalization, technology, and changing consumer perceptions. Today's retail sector is dominated by big retail chains like Walmart, Costco, and e-commerce giants like Amazon.[3]

American and European companies dominate the global retail market. Walmart Stores Inc. is the world largest retailer, and CVS and Amazon are the second and third largest, respectively. The other top global retailers include Costco Wholesale Corporation, Tesco PLC, Kroger's Corporation, Aldi Einkauf Gmbh & Co., OHG, Metro AG, The Home Depot, Inc., Careefour SA, and Target Corporation. Amazon is leading in e-commerce or online retailing with Alibaba being a close second.

In this chapter the key applications of analytics in the retail industry are discussed. We then discuss about the general nature of the ARIMA approach to the time series analysis and stages of the ARIMA model. We also discuss about the Seasonal ARIMA (SARIMA) model. We also present a case study on sales forecasting by using seasonal autoregressive integrated moving average models (SARIMA) in R and SAS Studio.

© Deepti Gupta 2018
D. Gupta, *Applied Analytics through Case Studies Using SAS and R*,
https://doi.org/10.1007/978-1-4842-3525-6_3

Supply Chain in the Retail Industry

The key players in the retail supply chain include manufacturers, wholesalers, retailers, and the consumers. Supply chain management is a critical aspect of any business or sector, but in the case of the retail sector, it is of the utmost importance. Success of today's retail giants heavily rely on optimizing the complex supply chain consisting of multiple national and international manufacturers and wholesalers. Efforts are also being made to reduce intermediaries to reduce the time from point of production to point of consumption as well as increase profit margins. Traditional retail supply chain sequences consisted of wholesalers bulk purchasing the products from the manufacturer and then reselling to various retailers. Retailers were responsible for managing final sales to customers. Retail giants like Walmart and Amazon are adopting creative supply chain management approaches; these include direct deals with manufacturers and having an optimized network of distribution centers as well as in some cases also managing their own transportation networks.[4] Figure 3-1 displays the traditional retail supply-chain key players and their roles.

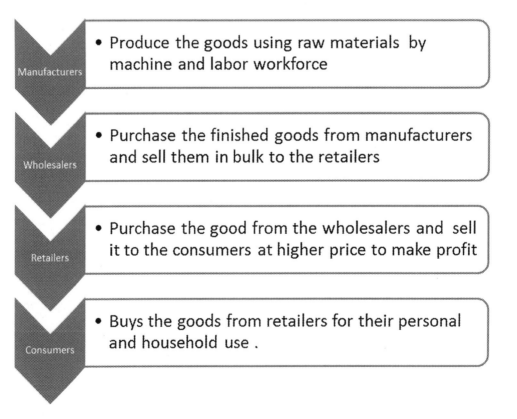

Figure 3-1. *Retail Supply chain*

Types of Retail Stores

There are several types of retail stores, each different from each other in terms of their business size, selling strategies, and pricing.[5]

- Department Stores: In department stores the end users can shop different products like apparel, electronics, houseware, etc., under one roof. Each department has its own marketing strategy to sell their products and serve the customers in the most efficient manner. Macy's Inc. is an example of a department store.

- Supermarket Stores: Supermarket retailers specialize in both food and non-food items at low and affordable prices. In supermarket stores the consumer can shop as well as avail other services, too, like pharmacy, flower shop, etc. Costco is an example of a supermarket.

- Convenience Stores: Convenience stores are mini-supermarkets and offer a limited range of product lines at premium prices as these stores are located in residential areas and open early and close very late at night. Seven Eleven (Seven & I Holdings Co.) is an example of a chain of convenience stores.

- Specialty Stores: In a specialty store, customers get specialized products specific to the particular industry. In such stores the customer service is very high as the salespeople there have the expertise and knowledge about their products. Specialty stores focus on store layout, merchandising, and branding to become popular. Zara Apparel is an example of a specialty store.

- Discount Stores: Discount stores are the type of retail outlets that offer a large variety of high discounted products by applying lot of tactics in order to increase the sales volume. Discount stores can be full-line discounters like Walmart, offering all types of goods at discounted rates such as apparel, housewares, toys, etc. Specialty discounters like Home Depot offer discounted rates on single lines.

- E-tailer: E-tailer retailers enable the customers to shop the products online and deliver the high-quality product at a reasonable price and fast delivery, like Amazon.

Role of Analytics in the Retail Sector

The retail industry is a consumer data-driven industry, where the bulk of consumer transactional data is generated on a daily basis. For example, Walmart is the biggest retailer in the world, which generates nearly one million customers' transactional data per hour, but collecting such diversified data is not enough for retailers. To withdraw the insights from such diversified data, predictive modeling and advance analytics are applied, which helps retailers to make effective decisions in today's volatile market.

Predictive analytics is widely used by both conventional retail stores as well as e=commerce firms for analyzing their historical data and building models for customer engagement, supply chain optimization, price optimization, and space optimization and assortment planning. An overview of some of these key applications is discussed in this section.

Customer Engagement

Predictive Analytics plays a key role in customer engagement. A vast amount of data is gathered by retailing firms through social media, website, in-store sales, etc. Success of the retailing industry heavily relies on understanding the consumer mindset and behavior and developing customer engagement strategies in order to attract more and more customers.

In today's highly competitive market, all retailers are replacing traditional blanket promotions to more sophisticated customer engagement approaches where retailers are targeting the consumer with personalized and customized offers.[6] To analyze a huge volume of online and in-store data, retailers are applying predictive modeling and machine learning algorithms to understand the 360-degree view of customers.[7] Customers' purchasing behavior is analyzed by pattern analysis models like association analysis or market basket analysis. These models help retailers to understand the likelihood of items or products, which tend to be purchased together.[8] Association rule mining helps in finding the correlation in a large set of data items by considering support, confidence, and lift in the Apriori algorithm.

Providing the customers a smart and satisfying shopping experience is always the value added in customer engagement. The recommendation engine and collaborative filtering technique is used to recommend the products to the consumer by observing their past purchasing behavior. Customer segmentation by using predictive modeling techniques like k-means is applied to segment their customers based on their purchasing behavior, likes, and dislikes.[9] All these techniques help retailers in knowing the customer behavior in better ways, tailor their marketing campaigns accordingly, and target them with the personalized and customized offers that result in more customers' purchases and loyalty.

Web and Social media analytics is another key driver to analyze the consumer mindset and helps in customer engagement. Analyzing consumer data and mining, customer reviews about the products and services from the social media like Twitter, Facebook and by using Natural language processing techniques,[10] text analytics, and sentiment analysis helps in understanding customer mindset in the effective way. For example, a retail company X would like to know about customers' thoughts and views about their products, so historical data of customers' reviews from online sales as well as in-store sales is analyzed by conducting sentiment analysis and scoring the sentiments of the consumers for their products and services. Whether the consumers have positive, negative, and neutral sentiments are helping the company understand the factors that are making customers happy and satisfied; and if this is not true, what are the factors that are making customers unhappy and unsatisfied. This will help the company strategize marketing and promotional approaches to resolve the issues of unhappy customers and regain their trust, hence increasing customer loyalty.[11] Happy customers are the key to increase business revenues and profits in retailing.

Supply Chain Optimization

Predictive Analytics and big data technologies are playing crucial a role in having effective and smooth supply chain management in the retailing business. Today's retail market is growing actively and is becoming highly competitive. In order to thrive in this competitive environment all the successful retailers like Walmart, CVS, and Costco are using their internal and external data to apply advance analytical and machine learning techniques to have effective supply chain management in order to drive sales and improve profit margins.

Supply chain is defined as the movement of the finished goods from the manufacturer to the supplier and finally delivered to the consumer. The success of effective supply chain management is to reduce the inventory and make the product available on shelves when required. Inventory management is the biggest challenge for all the retailers' and how to bridge the gap between demand and supply is necessary for all the retailers, as an excess of inventory as well as less inventory can damage the retailing business. Hence inventory optimization is necessary for every retailer to beat the competition and drive sales.

Effective logistic service is another critical factor, which is the matter of picking, packing, and shipping the finished goods. Modern logistic service helps in timely delivery of the goods and quicker replenishments stock to the stores. Automation devices like RFID technology, which uses electromagnetic fields to automatically identify and track tags attached to objects, is used in logistic companies to identify the location of the shipped items at any moment and how long shipped items will take to reach the destination.[12] By applying predictive analytics classification models like a decision tree, logistics regression, etc., in historical data can help the retailers determine whether the shipment of items will be delayed or not so the customers can be informed through emails.

Similarly, in the warehouse management system, RFID technology can help retailers find out all the details they would like to know about their box in less time and less prone to manual errors. Retailers using RFID technology even give their logistic center visibility through the shipping process.

An auto replenishment system helps retailers in tracking the out-of-stock rate and quicker replenishment of the stocks in the stores, like when to order, which quantity, etc., and for example, identification technologies (bar codes and scanners).[13] Such automation devices are used in the SCM industry in order to reduce the dependency on labor, which results in increase productivity and decrease in manual error.

Price Optimization

In retailing, pricing is a crucial factor to drive the gross margin and sales figures. Identifying the correct price for their products is extremely critical for retailers. High prices will erode customer loyalty, especially when the consumers are price sensitive and low prices or markdowns will generate low margins and losses to the retailers. Price optimization in the retail industry involves two main steps: price elasticity, which displays the impact of change in price for the demand of the products except luxury products; and price level optimization, which is all about setting the price that will maximize revenue of the products.

Traditional retailing used to rely solely on past experience and assumptions for product pricing, but now predictive analytics plays a powerful role in setting the right price for the right product at the right time. Every product is different, and all products are not sensitive to price. Identifying the factors that are responsible for driving the sales of the products helps the retailers to fix the optimum prices and create markdown strategies in a better way so unnecessary markdowns can be prevented, which results in millions of dollars of loss to the retailers.

Advance analytics method like a decision tree, logistic regression, demand model,[14] and price elasticity analysis can help in estimating the price response rates and fixing the optimum prices for each product.[15] This helps the retailers in generating more sales and revenue; similarly inter-store inventory can help in moving the products that are not getting sold in one store because of demographic factors in relation to the other stores, where the demand is high and it can be sold at profitable and full prices. These strategies are helpful in eliminating unnecessary markdowns and increasing profitability.

Space Optimization and Assortment Planning

In today's competitive world, space and assortment optimization has become a necessity for all retailers, manufacturers, and wholesalers in the retail industry. Assortment planning is the process about what product category, product width, and how much quantity should be present in the merchandise category based on size, style, color, etc., in order to drive high sales and high profits. Figure 3-2 displays the assortment planning process in the biggest retail giants like Walmart.

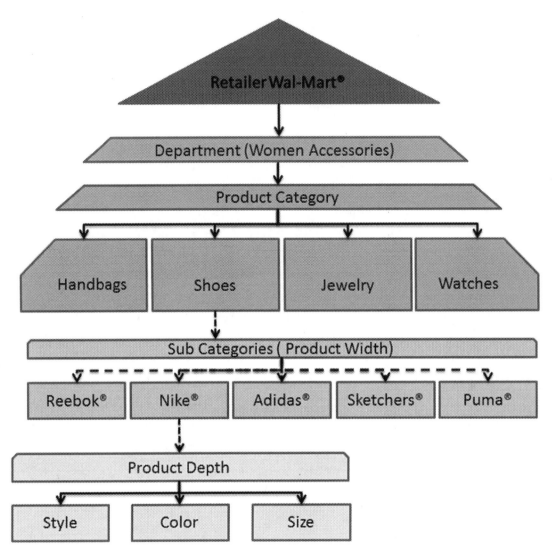

Figure 3-2. *Assortment planning process*

Efficient item assortment is helpful for all retailers both in terms of cost as well as in demand perspective. If the store has high-selling items, it minimizes the operation cost as well as improves customer satisfaction by providing them the most preferred SKU (stock keeping unit) and satisfies their needs. Inefficient items assortment raises the inventory cost. For example, if low-demand products are occupying the shelf for a long time. they are indirectly impacting the overall revenue potential of the store by blocking the possibility of positioning a high-selling item in its place. Also, poor, slow-moving assortments will give a stale look to the store, further impacting customer satisfaction.

Space optimization and assortment planning are inseparably linked. Which product category should be placed to what location in stores, how much space should be allocated to certain items or product category, is a very important decision for retailers, and it can vary from retailer to retailer based on the historical data. For example, the space allocated to bread in a convenience store will differ from the space allocated to the conventional grocery store as a convenience store may offer only 3 SKUs for bread while conventional grocery stores may offer 30 SKUs for bread, hence more shelf space. The bread is more important to the conventional grocery store and much less important to convenience store, so there is a difference in the space allocated to the bread and therefore the difference in the assortment. With the help of predictive analytics and machine learning algorithms, like a decision tree, market basket analysis,[16] etc., the historical data is analyzed and helps retailers to find out that which types of items or product categories should be located in which part of the store and which product category should be given more space on the shelf.

Case Study: Sales Forecasting for Gen Retailers with SARIMA Model

In this retail case study, the data is related to the sales of food and beverages of the retailer Glen. A machine learning algorithm, like ARIMA (Autoregressive integrated moving average) or ANN (Artificial neural network) multiple regression, etc.,[17] is used by retailers for sales forecasting, which help retailers in better planning of the inventory, stores' productivity, and effective business decisions.

Time series data is defined as the observed values in a series of regular time intervals, for example, monthly, weekly, quarterly, etc. If the data is not in a series of regular intervals, then the data is not fit for performing any forecasting. In such a scenario the gap in the data must be filled by converting the irregular data to a regular time series data. There are many ways to pad or aggregate the data and make the data fit for conducting a time series analysis. In a univariate time series, there is only one target variable and its values are used to forecast the future, and it is used in sales forecasting, weather forecasting, stock prices, etc. Time series data follow different types of patterns and are classified into trend, seasonal, cyclical, and random components.[18] Time series can have some or all of the four components in the data. These components are usually multiplied or added, that is,

$$Y_t = T+S+C+R \text{....... Additive}$$

$$Y_t = T^*S^*C^*R \text{..........Multiplicative}$$

1. Trend: Trend can be an upward or positive trend and downward or negative trend; it is a long-term pattern usually sustained for more than one year. When there is no increasing or decreasing trend, then it means the series is a stationary series in the mean and there is no need to apply differencing in the data.

2. Seasonal: A Pattern that appears in the data in a regular interval like high retail sales in December due to Christmas, and the pattern will repeat every year in December; the same is true for quarterly and daily data. Seasonality is always for a fixed period. Figure 3-3 displays the monthly retail sales data displaying seasonal patterns and it repeats every year in the month of December.

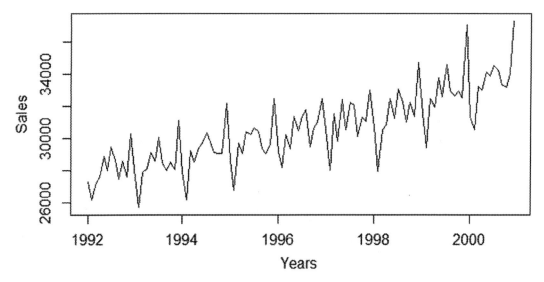

Figure 3-3. *Monthly retail sales data displaying seasonal pattern*

3. Cyclic: Cyclic pattern is a non-seasonal component and occurs in the data where there is rise and falls but not in a fixed period. Cyclic trends require at least more than one year, and cycles are not regular patterns and other than time, lots of other factors can affect cycles like economic recession, inflation, etc. For example, each country has a different budgeting policy at different time periods.

4. Random: A Random component is an unexplained and unpredictable component in the data. Every time series has some unpredictable and unexplained components. After all three series – trend, seasonality, and cyclic – are extracted from the series, the component that is left is random and unpredictable.

Overview of ARIMA Model

The ARIMA model is one of the widely used econometrics model in univariate time series data, and it was first used by Box and Jenkins for developing a univariate time series model for forecasting. ARIMA models are also known as Box-Jenkins models. ARIMA stands for autoregressive integrated moving average.[19] An ARIMA model has three parameters: AR, I, and MA. Where AR stands for autoregressive and is denoted by p, I stands for Integrated and is denoted by d, and MA stands for moving average and is denoted by q.

AutoRegressive Model

An AutoRegressive or an AR model is represented by parameter p. An autoregressive model is defined as the one where output variable Y_t is regressed on its own past values, for example:

Y_{t-1}, it represents AR = 1

Y_{t-2}, it represents AR=2

Y_{t-3} etc., it represents AR=3

A common representation of the AutoRegressive model is represented below.

$$Y_t = \beta 0 + \beta_1 Y_{t-1} + \beta_2 Y_{t-2} + \beta_3 Y_{t-3} + \cdots \beta_p Y_{t-p} + \varepsilon_t$$

Where

Y_t = y measured in time period t

$\beta 0$ = constant

$\beta_1 Y_{t-1}$ =AR (1) model, the value at time t is predicted from the values at times t−1

$\beta_2 Y_{t-2}$ =AR (2) model, the value at time t is predicted from the values at times t−1 and t-2

$\beta_3 Y_{t-3}$ = AR (3) model, the value at time t is predicted from the values at times t−1, t-2 and t-3

$\beta_p Y_{t-p}$ = AR (p) model, the value at time t is predicted from the values at times t-p

ε_t = Error term

Moving Average Model

Moving Average or an MA model is represented by parameter q. A moving average model or MA model is defined as where the output variable Y_t depends only on the random error terms of its own past values, for example:

ε_{t-1} it represents MA = 1

ε_{t-2}, it represents MA=2

ε_{t-3} etc., it represents MA=3

A common representation of the moving average model is represented below.

$$Y_t = \beta 0 + \varepsilon_t + \Phi_1 \varepsilon_{t-1} + \Phi_2 \varepsilon_{t-2} + \Phi_3 \varepsilon_{t-3} + \cdots \Phi_q \varepsilon_{t-q}$$

Where

Y_t = y measured in time period t

$\beta 0$ = constant

ε_t = error term

$\Phi_1 \varepsilon_{t-1}$ =MA (1) model, the value at time t depends on the past value of error term ε_{t-1}

$\Phi_2 \varepsilon_{t-2}$ = MA (2) model, the value at time t depends on the past value of error term ε_{t-1} and ε_{t-2}

$\Phi_3 \varepsilon_{t-3}$ = MA (3) model, the value at time t depends on the past value of error term ε_{t-1}, ε_{t-2} and ε_{t-3}

$\Phi_q \varepsilon_{t-q}$ = MA (q) model, the value at time t depends on the past value of error term ε_{t-q}

AutoRegressive Moving Average Model

When both AutoRegressive (AR) and Moving Average (MA) models are combined together, they become an ARMA model. In this situation, a time series model depends on p of its own past values and q of its past values of error. A common representation of an AutoRegressive moving average model is represented below.

$$Y_t = \beta 0 + \beta_1 Y_{t-1} + \beta_2 Y_{t-2} + \cdots \beta_p Y_{t-p} + \varepsilon_t + \Phi_1 \varepsilon_{t-1} + \Phi_2 \varepsilon_{t-2} + \cdots \Phi_q \varepsilon_{t-q}$$

The Integrated Model

In time series analysis, an autoregressive integrated moving average (ARIMA) model is a generalization of an autoregressive moving average (ARMA) model. Let's say we are interested in forecasting sales of food and beverages. Some errors will be associated with the forecasting. The ARMA model tells me that how my future sales of food and beverages are dependent on my previous sales of food and beverages and the errors associated with that. The ARIMA model also tells the same information but after accounting for the trend or season that exists in the sale of food and beverages. Thus, from an application point of view, ARMA models are not adequate for non-stationary time series whereas the ARIMA model can include the case of a non-stationary time series. The most important criteria in ARIMA model is that the series must be stationary. A series is considered to be stationary when Y does not depend on time; it signifies that the mean and variance of the series Y_t are time invariant as they do not change over time. Figure 3-4 displays a stationary series.

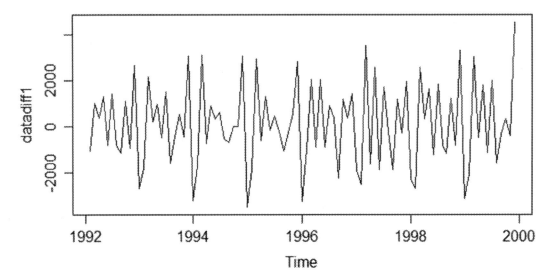

Figure 3-4. *Stationary series*

A series is considered to be non-stationary when Y depends on time and mean and variances do change over time. Figure 3-5 displays a non-stationary series.

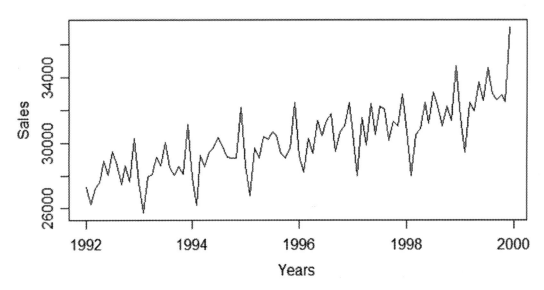

Figure 3-5. *Non-stationary series*

If the series is non-stationary, then the differencing is done to convert the series into being stationary, so a zero mean can be achieved and if the variance is not constant, then log transformation is done to make the variance constant. A stationary series will have a zero mean and constant variance. Differencing can be done one time or more to make the series stationary. When the series is differentiated one time, it is said to be integrated of order 1 and is denoted by I (1); a series differentiated two times is said to be integrated of order 2 and is denoted by I (2); and when the series is stationary without differencing, then it is represented as I (0) and in such situation, (d=0) and models are referred as an ARMA (p,q) model. Under differencing and over-differencing are not good because they can give the wrong prediction, so correct differencing is very essential to make a series stationary and to get the correct prediction.

Three Steps of ARIMA Modeling

The forecasting of univariate time series models using the Box-Jenkins approach is comprised of three stages: identification, estimation and forecasting. The Box-Jenkins approach is applicable only to stationary series, so the first step is to check whether the series is stationary. If the series is not stationary, then the help of differencing make the series stationary. The next step is to check for white noise, and if there is white noise in the model, then no ARIMA model is needed for the series. A detailed explanation of white noise and its impact is explained in program 1 of part 2.

Identification Stage

In the identification stage, differencing is done to make the series stationary. The Correlation approach is based on the Autocorrelation Factor (ACF) and Partial Autocorrelation Factor (PACF). Autocorrelation is defined as the correlation with itself or correlation with the same series, like correlation between Y_t and Y_{t-1}. ACF helps in identifying the number of lags for q. Partial Autocorrelation Function (PACF) is similar to ACF and is defined as the correlation between Y_t and Y_{t-p} when the correlation effect at the intermediate time lags are removed. PACF helps in identifying the number of lags for p.

The ACF correlogram is the plot of ACF against lag. and similarly the PACF correlogram is the plot of PACF against lag. With the help of ACF and PACF correlograms, the p (AR) and q (MA) values can be identified. Another approach is information criteria based on the maximum likelihood. The ARIMA (p,d,q) model having the lowest AIC value is considered as the best model. A detailed explanation of ACF and PACF with stationary and non-stationary series can be seen in the "Seasonal ARIMA in R and SAS" section of this chapter. Figure 3-6 displays the ACF correlogram, and Figure 3-7 displays the PACF correlogram.

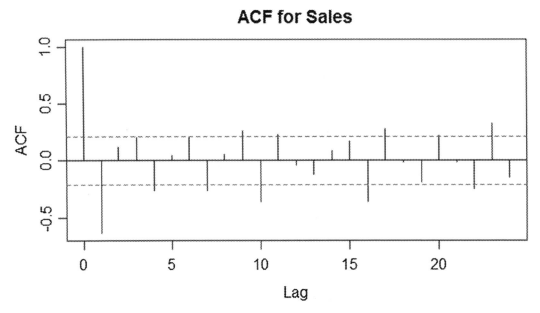

Figure 3-6. *ACF correlogram*

The Autocorrelation function (ACF) is a plot of correlation coefficients between a time series and lags of itself. By looking at the ACF plots, we can tentatively identify the numbers of AR terms.

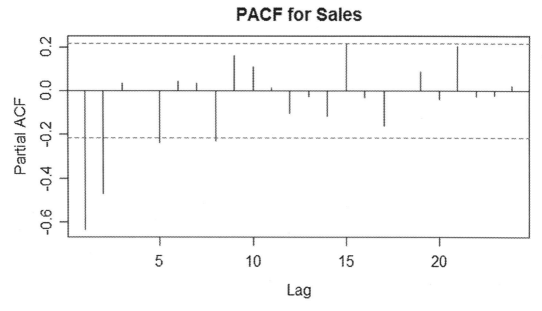

Figure 3-7. *PACF correlogram*

The Partial autocorrelation function (PACF) is a plot of partial correlation coefficients between a time series and lags of itself. By looking at the PACF plots, we can tentatively identify the numbers of MA terms.

Estimation and Diagnostic Checking Stage

In ARMA models, there are several methods for estimating the parameters of the models. They are Yule walkers, method of moments, conditional least squares, maximum likelihood method, etc., which are used for parameter estimation and it can be repeated till the best model is selected. In the diagnostic checking stage, the adequacy of the model is measured, with the help of significance tests; by looking at t values it indicates the significance and non-significance terms in the model. The goodness-of-fit statistics test helps in comparison of the models. Best model selection criteria are based on the lowest AIC and BIC values. Correlations of parameter estimates help in assessing the correlations between the parameter estimates. High correlations between parameter estimates can impact the results.

The Chi-Square test for residuals helps in identifying the white noise, and if there is autocorrelation in the residuals it means there is no white noise and if residuals are uncorrelated, then it signifies that there is white noise. Residuals must be uncorrelated and have a zero mean; if the residuals means are other than zero, it indicates that the forecasts are biased.

Forecasting Stage

The last stage is the forecasting stage where the predictions or future values of the time series are forecasted or predicted along with the standard error estimate and 95% confidence interval for forecasts, that is, a lower confidence interval and a higher confidence interval. Lead indicates that how many periods ahead to forecast like 12 months, 2 years, etc. ID indicates date time and uses to date the time series observations. Interval indicates whether data is monthly, weekly, yearly, etc., and the out option is used to store the output of the forecasts. Figure 3-8 displays forecasts for sales. The blue color line displays the forecasted values for sales.

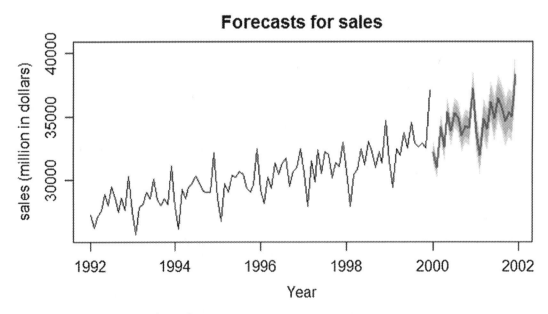

Figure 3-8. *Forecasts for sales*

Seasonal ARIMA Models or SARIMA

The seasonal ARIMA model or SARIMA includes both seasonal and non-seasonal factors that will be present in the multiplicative model. A seasonal ARIMA model is represented as:

$$ARIMA(p,d,q)\times(P,D,Q)s$$

Where

p = non-seasonal AutoRegressive or AR order

d = non-seasonal differencing (I) or Integrated order

q = non-seasonal Moving Average or MA order

P = seasonal AutoRegressive or AR order

D = seasonal differencing (I) or Integrated order

Q = seasonal Moving Average or MA order

S = time period of repeating seasonal pattern

For example, $ARIMA(1,1,1)\times(1,1,1)12$

The model includes a non-seasonal AR (1) term, a non-seasonal differencing of 1, a non-seasonal MA (1) term, a seasonal AR (1) term, a seasonal differencing of 1, a seasonal MA (1) term, and seasonal time period for monthly data is S = 12.

Seasonality in a time series is a continuous pattern that repeats over S time periods where S is the number of time periods of a repeating seasonal pattern. For example, when there is seasonality in monthly data, then S is equal to 12 time periods per year; when there is seasonality in quarterly data, then S is equal to 4 time periods per year; and when there is seasonality in weekly data, then S is equal to 7 time periods per year. Seasonality can result into a non-stationary series because of the repeating seasonal pattern in the data, for example, when there is seasonality in monthly sales data, the pattern of high and low sales repeats every year in the particular month.

Time series models involve lagged terms and may involve differenced data to account for trend. The Backshift operator B is a useful notational device for describing the process of differencing in time series lags. For example, Byt = yt−1 means that B operating on yt, has the effect of shifting the data back one time, similarly two applications of B to yt shifts the data back two times and so on.

115

To convert the series a from non-stationary to stationary series, seasonal differencing is done: for monthly data, seasonal differencing is equal to 12 and it can be represented as $\left(1-B^{12}\right)Y_t = Y_t - Y_{t-12}$. For quarterly data, seasonal differencing is equal to 4 and it can be represented as $\left(1-B^4\right)Y_t = Y_t - Y_{t-4}$. If in the data there is a trend, then non-seasonal differencing is applied, but more often non-seasonal differencing of 1 is applied and in some cases differencing can be 2, 3 also, depending upon the data .Non-seasonal differencing of 1 is represented as $\left(1-B\right)Y_t = Y_t - Y_{t-1}$. When in data there is both trend and seasonality, then to make the series stationary both non-seasonal differencing and seasonal differencing is applied. Figure 3-9 displays an ACF and PACF correlogram where both non-seasonal and seasonal factors are considered.

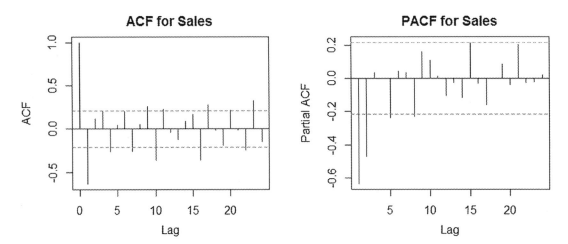

Figure 3-9. *ACF and PACF correlogram*

In the ACF and PACF correlogram in Figure 3-9, looking at the spikes of ACF at lags 0, 1, 10, and 16, then looking at the spikes of PACF at lags 1 and 2. ARIMA is represented as $ARIMA(2,1,1)\times(0,1,1)12$.

This model includes non-seasonal AR (2) term, non-seasonal differencing of (1), non-seasonal MA (1) term, seasonal AR (0) term, seasonal differencing of (1), seasonal MA (1) term, and seasonal time period for monthly data is S = 12.

Evaluating Predictive Accuracy of Time Series Model

There are many ways to measure the predictive accuracy. The common measures used for evaluating predictive accuracy of a time series model are scale-dependent measures like Mean absolute error (MAE), Root mean squared error (RMSE). and scale-independent measures like Mean absolute percentage error (MAPE).[20]

1. Mean absolute error (MAE): MAE is a popular measure for evaluating and predicting accuracy on a single data set of the same units. MAE is scale-dependent and it cannot be used to compare the series, which are on different units. For example, let x be the observed value and x^ is the forecast value, then the forecast error is the difference between (forecast – observed) value, which is e_i = x^ - x . Mean absolute error (MAE) is defined as the mean of all absolute errors and it is calculated as: MAE = mean (|x^ - x|).

    Where

    x^ = Forecast value
    x = Observed value

2. Root mean squared error (RMSE): RMSE is again a scale-dependent measure and is used to compare forecasts for the different models that are on the same scales or units. The smaller the error, the better is the predicting ability of the model. RMSE is a measure of how spread are the residuals or prediction errors and it is calculated as: RMSE = $\sqrt{mean\left(e_i^2\right)}$

3. Mean absolute percentage error (MAPE): MAPE is scale-independent measures so it is a frequently used method in comparing forecast for the different models having different scales or units. In MAPE, accuracy is displayed as percentage and it is calculated as:

$$\text{MAPE} = \frac{100}{N} \sum_{t=1}^{n} \left| x_t - x^{\wedge}/x_t \right|$$

Where

x_t = Actual values

$x^{\wedge}$ = Forecast values

$||$ = Absolute value symbol

The lower the MAPE value, the better is the predicting ability of the model and the higher the MAPE value, the worse is the forecast model. MAPE has some drawbacks as it is not applicable when there are zero values in the observed data. Very small observed values lead to the high percentage error. Another issue with percentage error is that scales are based on quantity, so it is useful only where the percentage remains the same with changing the scales.

Seasonal ARIMA Model Using R

In this Retail case study, we will discuss the data and variables used in the data. We then discuss about the exploratory data analysis in R, which is considered as the first step in the data analysis process. The exploratory data analysis helps in taking a broad look at patterns, trends, summary, outliers, missing values, and so on in the existing data by using a visual and quantitative method. We also discuss about building a Seasonal ARIMA model, model validation, and sales forecasting in R by using synthetic sales data of food and beverages.

Business Problem: Forecast retail food and beverages sales

Business Solution: Build the time series model using SARIMA

About Data

In this retail case study, seasonal the ARIMA model is built to forecast retail food and beverages sales and the data is generated synthetically. In this dataset there are a total of 309 observations and 2 variables, Period and Sales variables are numeric, and data is from January 1, 1992, to September 1, 2017. Sales is the dependent variable in the data. This historical data set is used to forecast retail food and beverages sales for the next 30 months.

In R, create your own working directory to import the dataset.

```
#Read the data from the working directory, create your own working directly
to read the dataset.

setwd("C:/Users/Deep/Desktop/data")

data1 <- read.csv ("C:/Users/Deep/Desktop/data/
final_sales.csv",header=TRUE,sep=",")

#Convert data into time series data

data <- ts(data1[,2],start = c(1992,1),frequency = 12)
data
```

Performing Data Exploration for Time Series Data

In Exploratory data analysis we are going to take a broad look at patterns, trends, seasonal differencing, non-seasonal differencing, summary, outliers, missing values, and so on in the existing data by using visual and quantitative method. R codes for data exploration and their output is discussed in the section below.

```
#Perform exploratory data analysis to know about the data

#Displays the start date of the time series data

start(data)
```

```
[1] 1992    1
```

```
# displays the end date of the time series data

end(data)

[1] 2017    9
```

```
#Displays the frequency of the time series data whether monthly, quaterly,
weekly.

frequency(data)

[1] 12
```

```
#Displays the data type it is time series data

class(data)
[1] "ts"
```

```
#Displays descriptive statistics of the time series data

 summary(data)
   Min    1st Qu   Median  Mean    3rd Qu   Max.
  28235   35168    42036   43133   50171    64663
```

```
#Checking the missing values present in the time series data

is.na(data)

 [1] FALSE FALSE FALSE FALSE FALSE FALSE FALSE FALSE FALSE FALSE
[11] FALSE FALSE FALSE FALSE FALSE FALSE FALSE FALSE FALSE FALSE
[21] FALSE FALSE FALSE FALSE FALSE FALSE FALSE FALSE FALSE FALSE
```

In this case partial output is displayed. False represents that there are no missing values present in the data; if there are missing values present in the data it will be represented as TRUE.

```
#Plotting the time series data

plot(data, xlab='Years', ylab = 'Sales')
```

Figure 3-10 displays monthly retail food and beverages sales data from 1992 to 2017. There is an upward trend in the data and the series is non-stationary with seasonality as sales (Y) depends on time and there is repetitive pattern of high sales in December of every year because of festivities and holidays. Converting the non-stationary series into stationary and taking the seasonal differencing is the first important step in the ARIMA model. The stationary series will have a zero mean and constant variance.

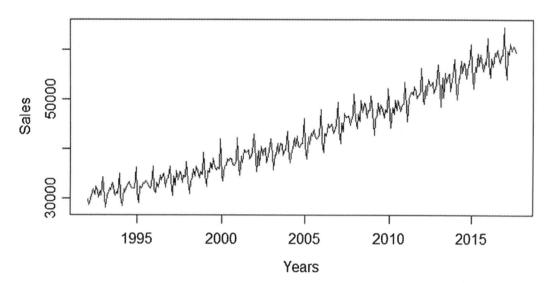

Figure 3-10. *Monthly retail food and beverages sales from 1992–2017*

```
#Install astsa package

install.packages("astsa")
library(astsa)

#To see acf and pacf in original data

acf2(data,max.lag = 24)
```

```
        ACF    PACF
 [1,]  0.95   0.95
 [2,]  0.93   0.30
 [3,]  0.92   0.23
 [4,]  0.92   0.16
 [5,]  0.92   0.24
 [6,]  0.90  -0.19
 [7,]  0.90   0.25
 [8,]  0.89  -0.21
 [9,]  0.87   0.01
```

```
[10,] 0.86 -0.20
[11,] 0.86  0.32
[12,] 0.89  0.28
[13,] 0.84 -0.56
[14,] 0.82 -0.01
[15,] 0.81  0.12
[16,] 0.81  0.01
[17,] 0.81  0.01
[18,] 0.79  0.03
[19,] 0.80  0.07
[20,] 0.78 -0.17
[21,] 0.76  0.00
[22,] 0.75  0.15
[23,] 0.75  0.05
[24,] 0.77 -0.04
```

Figure 3-11 displays the autocorrelation function (ACF) and partial autocorrelation function (PACF) in the original data, which is the non-stationary series. ACF and PACF graphs decay very slowly, which signify that the series is non-stationary.

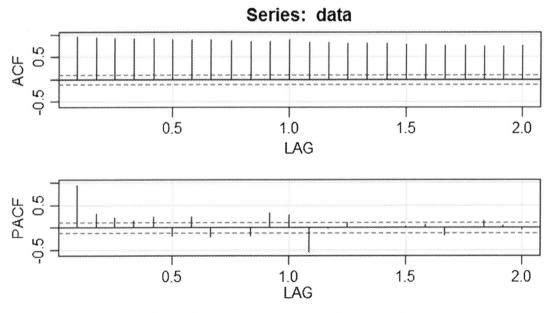

Figure 3-11. *ACF and PACF representing non-stationary series*

```
#Seasonally differenced retail sales
datadiff12 <- diff(data,12)
#Plot seasonally differenced retail sales
 plot.ts(datadiff12)
```

Figure 3-12 displays the seasonally differenced data as it is monthly retail sales data so 12 is considered, if data is quarterly it is 4, for weekly it is 7, etc.

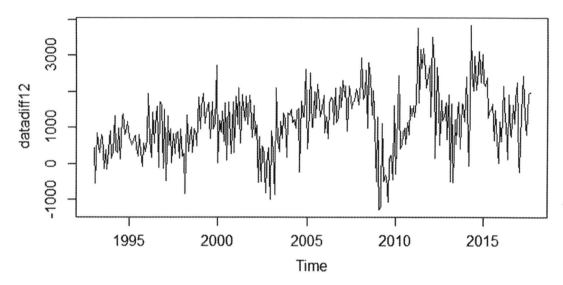

Figure 3-12. *Seasonally differenced retail sales*

```
#To see acf and pacf of seasonally differenced retail sales

acf2(datadiff12,max.lag = 24)
```

```
       ACF  PACF
[1,]  0.33  0.33
[2,]  0.45  0.39
[3,]  0.55  0.44
[4,]  0.31  0.02
[5,]  0.42  0.07
[6,]  0.40  0.08
[7,]  0.16 -0.22
[8,]  0.27 -0.12
```

```
[9,]   0.27  0.09
[10,]   0.04 -0.14
[11,]   0.15 -0.09
[12,]  -0.03 -0.21
[13,]  -0.02  0.01
[14,]   0.08  0.13
[15,]  -0.02  0.16
[16,]  -0.11 -0.06
[17,]   0.03  0.07
[18,]  -0.05  0.08
[19,]  -0.11 -0.12
[20,]   0.01 -0.04
[21,]  -0.12  0.00
[22,]  -0.14 -0.18
[23,]   0.07  0.10
[24,]  -0.15 -0.06
```

Figure 3-13 displays the ACF and PACF plot, which decays quickly.

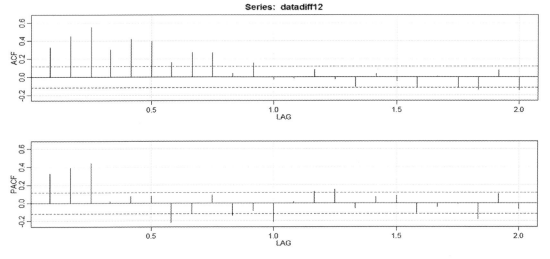

Figure 3-13. *Seasonally differenced ACF and PACF plot*

```
#Trend and seasonally differenced retail sales

diff1and12=diff(datadiff12,1)

#Plot Trend and seasonally differenced retail sales

plot(diff1and12)
```

Figure 3-14 displays the trend and seasonally differenced retail sales by considering an additional first difference.

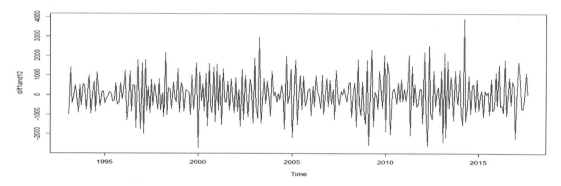

Figure 3-14. *Trend and seasonally differenced retail sales*

```
#To see acf and pacf of trend and seasonally differenced retail sales

acf2(diff1and12,max.lag = 36)
```

```
         ACF   PACF
[1,]   -0.60  -0.60
[2,]    0.03  -0.52
[3,]    0.25  -0.06
[4,]   -0.27  -0.11
[5,]    0.10  -0.12
[6,]    0.17   0.18
[7,]   -0.26   0.08
[8,]    0.08  -0.12
[9,]    0.17   0.09
```

```
[10,] -0.26  0.02
[11,]  0.22  0.13
[12,] -0.15 -0.10
[13,] -0.06 -0.20
[14,]  0.15 -0.20
[15,] -0.02  0.03
[16,] -0.16 -0.11
[17,]  0.17 -0.12
[18,] -0.02  0.09
[19,] -0.14  0.01
[20,]  0.20 -0.03
[21,] -0.09  0.14
[22,] -0.17 -0.13
[23,]  0.32  0.03
[24,] -0.20  0.02
[25,] -0.12 -0.27
[26,]  0.28 -0.23
[27,] -0.19  0.00
[28,]  0.01 -0.02
[29,]  0.13 -0.14
[30,] -0.16  0.02
[31,]  0.02 -0.01
[32,]  0.15 -0.05
[33,] -0.17  0.12
[34,]  0.01 -0.15
[35,]  0.20  0.12
[36,] -0.32 -0.08
```

Figure 3-15 displays the ACF and PACF plot.

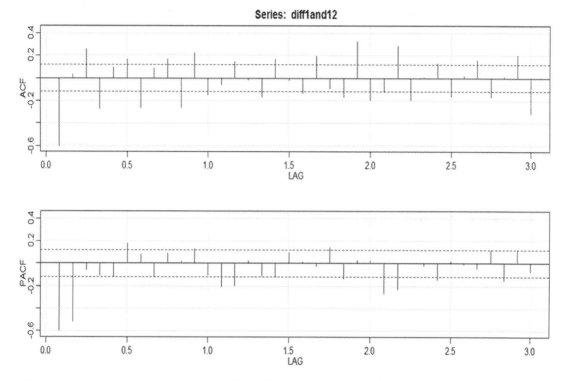

Figure 3-15. *Trend and seasonally differenced ACF and PACF plot*

The next step is to identify the appropriate SARIMA model based on the ACF and PACF shown in Figure 3-15 Looking at the spikes of ACF, the significant spikes at lag 1 in the ACF signify a non-seasonal MA (1) component and the significant spikes at lag 12 and 24 signify a seasonal MA (2) component. Similarly, looking at the spikes of PACF the significant spikes at lag 2 in the PACF signify a non-seasonal AR (2) component and the significant spikes at lag 12 and 24 signify a seasonal AR (2) component. As a result, let's begin with an $ARIMA(2, 1, 1)(2, 1, 2)_{12}$ model, which indicates a non-seasonal and seasonal AR (2) component, non-seasonal MA(1) and seasonal MA(2) component, first and seasonal difference.

```
#Install forecast package

install.packages("forecast")
library(forecast)
#Building seasonal ARIMA(2,1,1)(2,1,2)₁₂    model
```

```
model1<- arima(data,order=c(2,1,1),seasonal=list
(order=c(2,1,2),period=12))
```

```
summary(model1)
Call:
arima(x = data, order = c(2, 1, 1), seasonal = list
(order = c(2, 1, 2), period = 12))
```

```
Coefficients:
          ar1       ar2      ma1     sar1     sar2     sma1     sma2
      -0.5122   -0.2391  -0.4420   0.8459  -0.5203  -1.6017   0.8261
s.e.   0.1175    0.0960   0.1073   0.0764   0.0686   0.0825   0.0835
```

```
sigma^2 estimated as 288663:log likelihood = -2295.91 aic = 4607.83
```

```
Training set error measures:
                  ME      RMSE      MAE        MPE       MAPE       MASE
Training set 37.34187 525.8883 384.6296 0.07293464 0.8940933 0.1821944
                ACF1
Training set 0.003230019
```

There are several error measures and most commonly used error measures are mean absolute error(MAE), root mean squared error(RMSE), mean absolute percentage error(MAPE), and akaike's information criteria (AIC) value is **4607.83**

```
# Portmanteau or Box-Ljung test to check whether residual are white noise
```

```
Acf(residuals(model1))
```

The residual for the $ARIMA(2, 1, 1)(2, 1, 2)_{12}$ model1 is displayed in Figure 3-16.

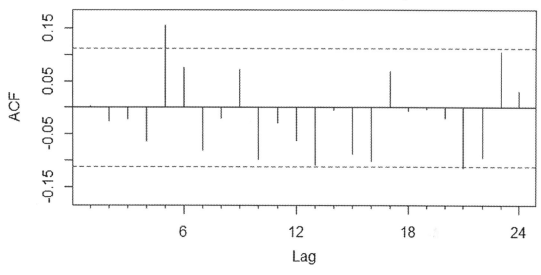

Figure 3-16. *ACF Residual from the ARIMA(2, 1, 1)(2, 1, 2)$_{12}$ model for retail sales*

```
# Portmanteau or Box-Ljung test to check whether residual are white noise

Box.test(residuals(model1),lag=24,fitdf =1,type="Ljung")

Box-Ljung test

data:  residuals(model1)
X-squared = 42.338, df = 23, p-value = 0.008288
```

Looking at ACF residuals spike at lag 5 are considered a significant spike as they are crossing the boundary line and it signifies that there is autocorrelation in the residuals, hence no white noise and some additional non-seasonal terms need to be added in the model. The Box-Ljung test also signifies that the residuals are autocorrelated as the p value is less than 0.05.

Let's try few other ARIMA models with different non-seasonal terms to identify best models like ARIMA(3,1,2)(2,1,2)$_{12}$, ARIMA(4,1,2)(2,1,2)$_{12}$, ARIMA(5,1,2)(2,1,2)$_{12}$ ARIMA(3,1,1)(2,1,2)$_{12}$ but none of the models are having a lower AIC value and able to clear all the necessary residual checks. ARIMA(6,1,1)(2,1,2)$_{12}$model is the only model that is having a lower AIC value as compared to other models and is able to pass all the necessary residual checks; hence ARIMA(6,1,1)(2,1,2)$_{12}$ model will be considered the final model for forecasting the retail sales.

#Rebuilding $ARIMA(6,1,1)(2,1,2)_{12}$ model with different non seasonal terms

```
model2 <- arima(data,order=c(6,1,1),seasonal= list
(order=c(2,1,2),period=12))
```

```
summary(model2)
Call:
arima(x = data, order = c(6, 1, 1), seasonal = list(order = c(2, 1, 2),
period = 12))
```

Coefficients:
```
  ar1     ar2      ar3      ar4      ar5     ar6       ma1     sar1
-0.5080 -0.2343 -0.0088 -0.0712 0.1201 0.1771   -0.4430  0.8258
```

s.e.
```
0.2929 0.2795  0.2047  0.1095  0.0959  0.0626   0.2966  0.0849
```

```
sar2      sma1     sma2
-0.4703 -1.5872 0.7996
```

s.e.
```
0.0764   0.0836   0.0859
```

sigma^2 estimated as 277602: log likelihood = -2288.44,
aic = 4600.89

Training set error measures:
```
       ME     RMSE      MAE       MPE       MAPE       MASE
Training set:
    33.64523 515.7163 376.6665 0.06388241 0.8774594 0.1784224
                ACF1
Training set -0.008903652
```

$ARIMA(6,1,1)(2,1,2)_{12}$ model, error values of all the measure and AIC values are lower than the other models, and akaike's information criteria (AIC) value for this model is **4600.89**.

```
# Portmanteau or Box-Ljung test on model2 to check whether residual are
white noise
```

```
Acf(residuals(model2))
```

The residual for the $ARIMA(6,1,1)(2,1,2)_{12}$ model1 is displayed in Figure 3-17.

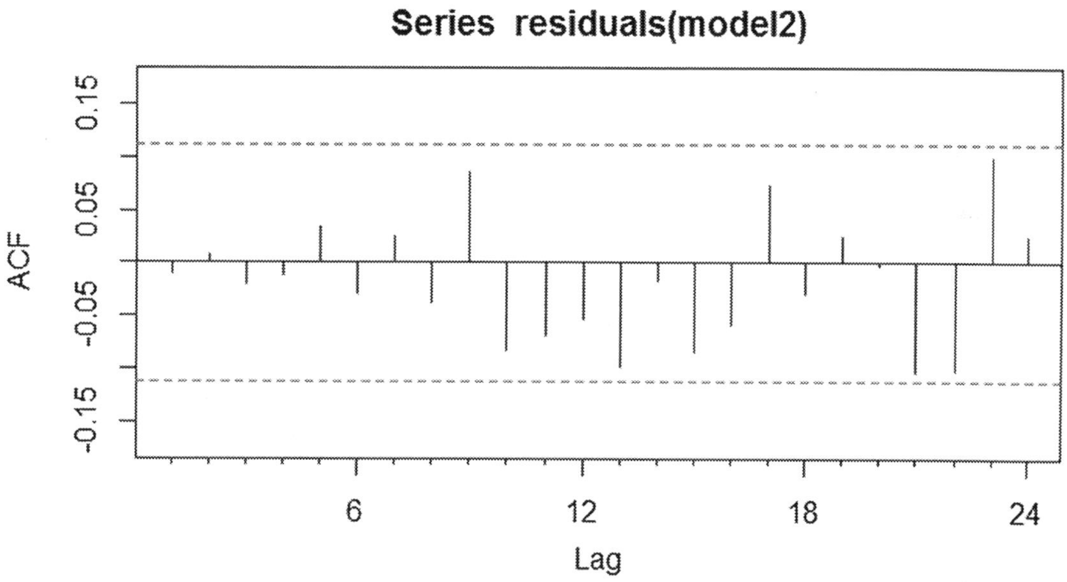

Series residuals(model2)

Figure 3-17. *ACF residuals from the $ARIMA(6,1,1)(2,1,2)_{12}$ model for retail sales*

```
# Portmanteau or Box-Ljung test on model2 to check whether residual are
white noise
```

```
Box.test(residuals(model2),lag=24,fitdf = 1,type="Ljung")
```

```
Box-Ljung test
data:  residuals(model2)
X-squared = 28.093, df = 23, p-value = 0.2123
```

Looking at ACF residuals, all of the spikes are within significance limits as none of the spikes is crossing the boundary line and it signifies that there is no autocorrelation in the residuals; hence there is white noise. The Box-Ljung test also signifies that the residuals are not autocorrelated as the p value is greater than 0.05. $ARIMA(6,1,1)(2,1,2)_{12}$ model cleared all the necessary residual checks and is ready for forecasting.

```
#Forecast for the next 30 month

Pred <- forecast(model2,h=30)
Pred
```

	Point Forecast	Lo 80	Hi 80	Lo 95	Hi 95
Oct 2017	60284.77	59609.22	60960.33	59251.60	61317.95
Nov 2017	60758.37	60082.00	61434.74	59723.96	61792.78
Dec 2017	65222.42	64516.77	65928.07	64143.22	66301.62
Jan 2018	59425.75	58673.63	60177.86	58275.48	60576.01
Feb 2018	56573.13	55806.50	57339.77	55400.66	57745.60
Mar 2018	61788.31	60958.80	62617.83	60519.68	63056.95
Apr 2018	60247.48	59369.55	61125.41	58904.80	61590.16
May 2018	63488.29	62594.77	64381.80	62121.78	64854.80
Jun 2018	61402.94	60467.17	62338.70	59971.81	62834.06
Jul 2018	62857.83	61893.03	63822.64	61382.29	64333.38
Aug 2018	63056.52	62063.70	64049.34	61538.13	64574.91
Sep 2018	60366.39	59335.22	61397.56	58789.35	61943.43
Oct 2018	62188.53	61085.91	63291.16	60502.21	63874.86
Nov 2018	62662.96	61533.08	63792.84	60934.96	64390.96
Dec 2018	66598.80	65427.76	67769.85	64807.84	68389.76
Jan 2019	61790.72	60582.20	62999.23	59942.45	63638.98
Feb 2019	58463.35	57221.13	59705.57	56563.54	60363.16
Mar 2019	63495.57	62211.24	64779.90	61531.35	65459.78
Apr 2019	61628.92	60308.33	62949.52	59609.25	63648.60
May 2019	65356.65	64004.49	66708.81	63288.69	67424.60
Jun 2019	63152.29	61764.24	64540.35	61029.44	65275.14
Jul 2019	64876.45	63456.29	66296.61	62704.50	67048.40
Aug 2019	64813.33	63360.62	66266.04	62591.60	67035.06
Sep 2019	61785.53	60299.75	63271.31	59513.23	64057.84
Oct 2019	64033.38	62525.08	65541.67	61726.64	66340.12
Nov 2019	64329.31	62790.78	65867.84	61976.33	66682.29
Dec 2019	68564.59	66997.95	70131.24	66168.62	70960.57
Jan 2020	63661.20	62068.37	65254.03	61225.17	66097.23
Feb 2020	60003.12	58382.56	61623.68	57524.69	62481.55
Mar 2020	65104.54	63458.84	66750.23	62587.66	67621.41

```
#Creating the plot for forecast retail sales

plot(Pred,ylab="sales (million in dollars)",xlab="Year")
```

Figure 3-18 displays the forecast of the retail sales data using $ARIMA(6,1,1)(2,1,2)_{12}$ for the next 30 months.

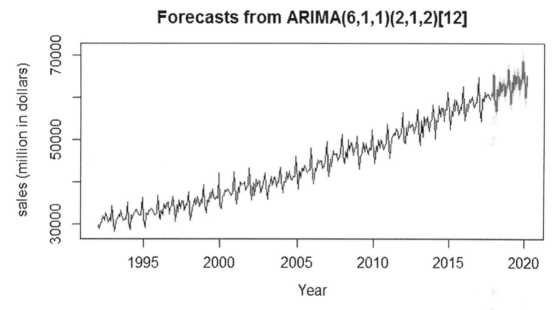

Figure 3-18. *Forecasts of the retail sales data using $ARIMA(6,1,1)(2,1,2)_{12}$ model*

Seasonal ARIMA Model Using SAS

In this section, we will discuss about the SAS procedures like proc content, checking for non-stationarity by Augmented Dicky Fuller, white noise, Conversion of non-stationary series into stationary by trend and seasonal differencing, ACF and PACF curve. We also discuss about building a Seasonal ARIMA mode in SAS for sales forecasting, with an explanation of SAS codes and output of each part in program 1 and program 2.

```
/*Create your own library in SAS like here it is libref and mention the
path of your data*/

libname libref "/home/aro1260/deep";

/* Importing final_sales dataset */

PROC IMPORT DATAFILE= "/home/aro1260/data/final_sales.csv"
DBMS=CSV Replace
OUT=libref.sales;
GETNAMES=YES;
RUN;

/* To check the contents of the data */

    PROC CONTENTS DATA = libref.sales;
    RUN;
```

The CONTENTS Procedure

Data Set Name	LIBREF.SALES	Observations	309
Member Type	DATA	Variables	2
Engine	V9	Indexes	0
Created	11/06/2017 14:35:16	Observation Length	16
Last Modified	11/06/2017 14:35:16	Deleted Observations	0
Protection		Compressed	NO
Data Set Type		Sorted	NO
Label			
Data Representation	SOLARIS_X86_64, LINUX_X86_64, ALPHA_TRU64, LINUX_IA64		
Encoding	utf-8 Unicode (UTF-8)		

Alphabetic List of Variables and Attributes

#	Variable	Type	Len	Format	Informat
1	Period	Num	8	DATE8.	DATE8.
2	Sales	Num	8	BEST12.	BEST32.

In this case, partial output of Proc contents is displayed. Proc contents displays the content of the data like there are total 309 observations, 2 variables, and both the variables are numeric.

```
/*plotting time series */
   proc timeseries data = libref.sales plot=series;
   id Period interval = month;
   var Sales;
   run;
```

The TIMESERIES Procedure

Input Data Set

Name	LIBREF. SALES
Label	
Time ID Variable	Period
Time Interval	MONTH
Length of Seasonal Cycle	12

Variable Information

Name	Sales
Label	
First	JAN1992
Last	SEP2017
Number of Observations Read	309

The time series procedure displays the information about the data like start data is from January 1992 and the end date is on September 2017 and the number of observations used is 309. The time series plot of the retail sales is shown in Figure 3-19. From the plot, it is clearly visible that the retail sales are having an increasing trend and the hikes in the peak at a constant time interval signify the presence of seasonality.

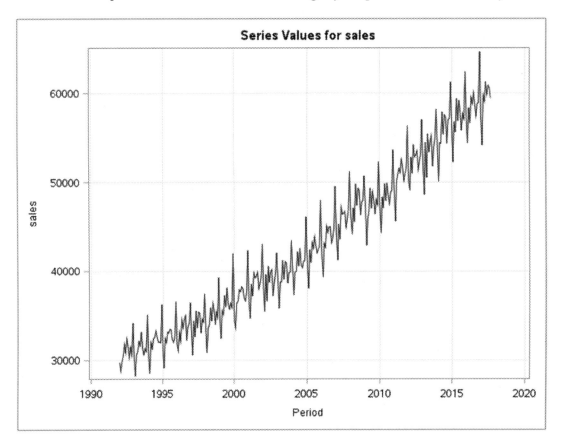

Figure 3-19. *Time series plot of retail sales*

```
/*checking for non stationarity by Augmented Dicky Fuller
 Test and looking at Auto correlation function */
```

In this section, non-stationarity of the data is checked by using the Augmented Dicky Fuller test and partial output is displayed. Complete output of the model with an explanation of each part will be discussed in program 1 and program 2.

```
proc arima data = libref.sales;
identify var =sales stationarity= (adf);
run;
```

The ARIMA Procedure

Name of Variable = sales

Mean of Working Series	43132.54
Standard Deviation	9091.094
Number of Observations	309

Augmented Dickey-Fuller Unit Root Tests

Type	Lags	Rho	Pr < Rho	Tau	Pr < Tau	F	Pr > F
Zero Mean	0	0.1189	0.7103	0.11	0.7178		
	1	0.4589	0.7959	0.65	0.8567		
	2	0.5853	0.8275	1.19	0.9404		
Single Mean	0	-12.7522	0.0671	-2.49	0.1188	3.31	0.2239
	1	-5.4862	0.3882	-1.58	0.4931	1.74	0.6252
	2	-2.3233	0.7388	-0.95	0.7725	1.45	0.7010
Trend	0	-229.770	0.0001	-13.48	<.0001	90.90	0.0010
	1	-213.903	0.0001	-10.27	<.0001	52.72	0.0010
	2	-158.255	0.0001	-7.85	<.0001	30.86	0.0010

In this case partial output is displayed. Looking at the Augmented Dickey-Fuller Tests and checking Tau statistics as part of the table (**Pr < Tau),** the values of (**Pr < Tau),** should be less than 0.05, which is not the case here; hence the series is not stationary. Figure 3-20 displays the Autocorrelation function chart and it is clearly visible that the ACF curve decays very slowly, indicating that the series is non-stationary.

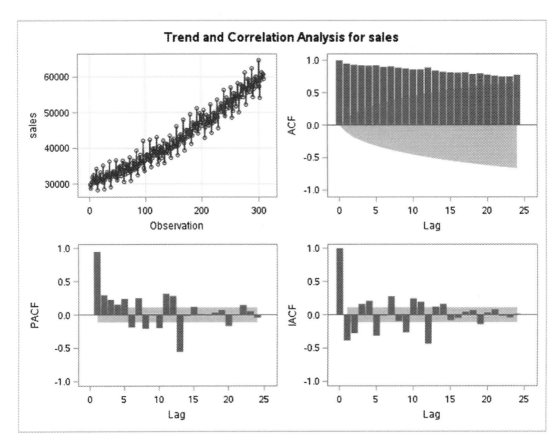

Figure 3-20. *Non-stationary series of retail sales*

```
/*Converting series into stationary by trend and seasonal differencing */
  proc arima data = libref.sales;
  identify var = sales(1,12) stationarity = (adf) ;
  run;
```

The ARIMA Procedure

Name of Variable = sales	
Period(s) of Differencing	1,12
Mean of Working Series	5.222973
Standard Deviation	1050.084
Number of Observations	296
Observation(s) eliminated by differencing	13

Conversion of non-stationary to stationary series is achieved by differencing both trend (1) and seasonal differencing (12) as retail sales are monthly data.

Augmented Dickey-Fuller Unit Root Tests

Type	Lags	Rho	Pr < Rho	Tau	Pr < Tau	F	Pr > F
Zero Mean	0	-471.802	0.0001	-34.34	<.0001		
	1	-1477.16	0.0001	-27.06	<.0001		
	2	-2523.29	0.0001	-15.28	<.0001		
Single Mean	0	-471.818	0.0001	-34.28	<.0001	587.63	0.0010
	1	-1477.66	0.0001	-27.02	<.0001	364.99	0.0010
	2	-2529.37	0.0001	-15.26	<.0001	116.41	0.0010
Trend	0	-471.817	0.0001	-34.22	<.0001	585.65	0.0010
	1	-1477.95	0.0001	-26.97	<.0001	363.76	0.0010
	2	-2533.02	0.0001	-15.23	<.0001	116.01	0.0010

Looking at the Augmented Dickey-Fuller Tests and checking Tau statistics as part of the table (**Pr < Tau**), the values of (**Pr < Tau**), is less than 0.05; hence the series is stationary. Figure 3-21 displays the trend and Correlation Analysis chart and it is clearly visible that the trend is constant in term of its mean and variance and ACF curve decays quickly, indicating that the series is stationary for retail sales.

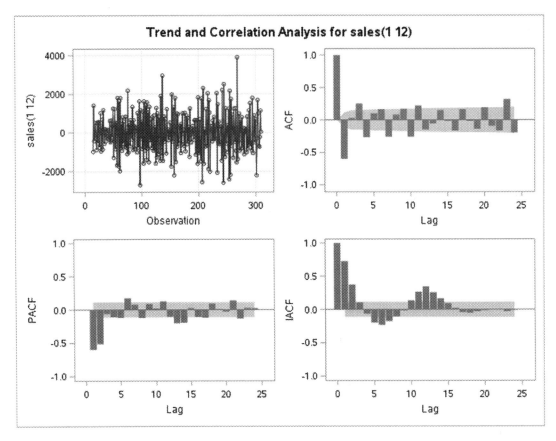

Figure 3-21. *Stationary series of retail sales*

```
/*Fitting seasonal ARIMA model for the retail sales  */
 /*by the maximum likelihood method  */
```

Program1:

```
proc arima data = libref.sales;
identify var = sales(1,12) stationarity= (adf) ;
estimate p = (2)(12) q = (1)(12)  method = ml;
```

```
/*Forecasting next 30 months  */
```

```
forecast lead = 30 interval = month id = Period
printall out = libref.pred_results;
run;
```

In the above program1 code **Proc arima** is the procedure used to build the forecasting model. The **Identify** statement is used to read the sales series, analyzing its correlation properties and analyzing **trend (1)** and **seasonal differencing (12)** as the data is monthly. Differencing is done to convert the non-stationary into stationary series, which is the critical step in ARIMA modeling. **Stationarity = (adf)** is used to display the results of Augmented Dickey Fullers Test, which is the check for a stationary series.

In the **estimation** of an ARMA model, the order of the autoregressive model with the non-seasonal and seasonal **p = (2) (12)** and the order of the moving average model with the non-seasonal and seasonal q = **(1) (12)** is mentioned. The **Estimate** statement is used to display the parameter estimates tables and various diagnostic or goodness-of-fit statistics, which signify how well the model is fitting the data. **In this forecasting model Method = ml** is specifying the estimation method which is maximum likelihood method in this case other estimation methods used are conditional least square method and unconditional least square method.

The **Forecast** statement is used to forecast or predict, **lead = 30** indicates **that** how many periods ahead to predict and in this case study the next 30 months of retail sales are predicted; **interval = month** indicates that there is monthly data; **id = Period**, which is a datetime variable and used to date the observations of the retail Sales; **out = libref.pred_results** create the forecast output and save the output in the **library libref as pred_results** dataset.

The **Seasonal ARIMA output of program1** is split into several sections and each is discussed below.

Part 1 of program1 displays the descriptive statistics for the retail sales series like period of differencing, mean of working series, standard deviation, number of observations, and number of observations eliminated by differencing.

Part 1

The ARIMA Procedure	
Name of Variable = sales	
Period(s) of Differencing	1,12
Mean of Working Series	5.222973
Standard Deviation	1050.084
Number of Observations	296
Observation(s) eliminated by differencing	13

Part 2 of program1 displays the table of the autocorrelation check for white noise and the autocorrelations are displayed in groups of 6, depending upon the humber of lags, by default the number of lags is 24. To build the ARIMA model the values must be autocorrelated with its past values; if the values are not autocorrelated, then there is no point of building an ARIMA model. From the table it is clearly visible that all the p values in the respective lags are less than 0.05, indicating that there are autocorrelations, so hence there is no white noise.

Part 2

Autocorrelation Check for White Noise

To Lag	Chi-Square	DF	Pr > ChiSq	Autocorrelations					
6	160.50	6	<.0001	-0.599	0.029	0.254	-0.270	0.099	0.167
12	235.79	12	<.0001	-0.261	0.083	0.171	-0.261	0.225	-0.148
18	261.18	18	<.0001	-0.058	0.149	-0.017	-0.165	0.166	-0.018
24	337.27	24	<.0001	-0.137	0.196	-0.092	-0.165	0.323	-0.197

Figure 3-22 is shown in Part 3 of program1 and it displays the stationary time series plot, autocorrelation function plot (ACF), partial autocorrelation function plot (PACF), and inverse autocorrelation function plot (IACF). In ARIMA modeling the inverse autocorrelation plot (IACF) almost plays the same role as partial autocorrelation function (PACF), but it generally indicates subset and seasonal autoregressive models better than the partial autocorrelation function (PACF). In addition, the inverse autocorrelation plot (IACF) is also useful in identifying the over-differencing in the data. These plots are used for trends and its autocorrelation analysis. Looking at these plots, whether the series is stationary or non-stationary can be identified. From the ACF plot it is clearly visible that the ACF plot decays very quickly, indicating that the series is stationary.

Part 3

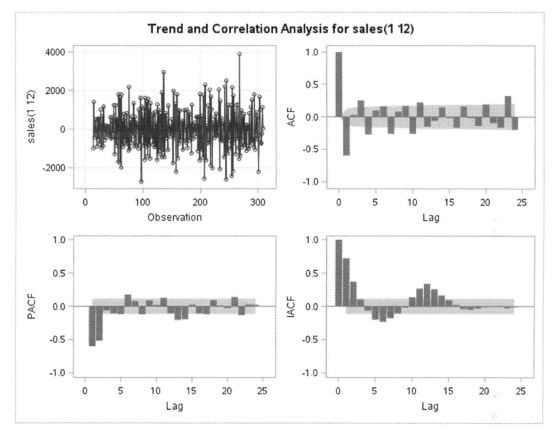

Figure 3-22. *Program1 Stationary time series*

Part 4 of program1 displays results of the maximum likelihood estimation table. The mean term is represented as MU and its estimated value is 3.82242. The moving average term is represented as MA1,1 estimated value is 0.78801 and MA2,1 estimated value is 0.87869. The autoregressive term is represented as AR1,1 estimated value is 0.07605 and AR2,1 estimated value is 0.23591. The t values and p values are considered as significance tests for parameter estimates and helps in identifying the insignificant and unnecessary terms in the model. In this case (MA1,1), (MA2,1), (AR2,1) terms are considered as significant as t values are >2 and p values are less than 0.05 whereas MU and AR1,1 are considered insignificant terms as t values are not >2 and p values are not less than 0.05.

Part 4

Maximum Likelihood Estimation

Parameter	Estimate	Standard Error	t Value	Approx Pr > ltl	Lag
MU	3.82242	1.99095	1.92	0.0549	0
MA1,1	0.78801	0.03940	20.00	<.0001	1
MA2,1	0.87869	0.05438	16.16	<.0001	12
AR1,1	0.07605	0.06552	1.16	0.2458	2
AR2,1	0.23591	0.07997	2.95	0.0032	12

Part 5 of program1 displays a goodness-of-fit statistics table, which is used in comparing the models and selecting the best model. The Constant estimate is the function of the mean term MU, moving average and autoregressive term. The Variance estimate is the residual series variance, and Std error estimate is the square root of the variance. AIC and SBC statistics are used to compare the models, and lower values of AIC and SBC indicate the better fitting model. In this model AIC value is 4692.91 and SBC value is 4711.362.

Part 5

Constant Estimate	2.698552
Variance Estimate	422169.6
Std Error Estimate	649.7458
AIC	4692.91
SBC	4711.362
Number of Residuals	296

Part 6 of program1 displays the correlations of parameters estimates table. In case two parametes are highly correlated with each other, one parameter must be dropped as it might impacts the results. In this case none of the parameters is highly correlated with each other.

Part 6

Correlations of Parameter Estimates

Parameter	MU	MA1,1	MA2,1	AR1,1	AR2,1
MU	1.000	-0.014	0.057	0.011	0.036
MA1,1	-0.014	1.000	0.062	0.419	0.187
MA2,1	0.057	0.062	1.000	0.053	0.616
AR1,1	0.011	0.419	0.053	1.000	0.277
AR2,1	0.036	0.187	0.616	0.277	1.000

Part 7 of program1 displays an autocorrelation check of a residuals table. Residuals are considered white noise when they are uncorrelated. In this case residuals are autocorrrelated, which means there is no white noise; hence the model is inadequate. Forecasting can be done for this model also, but the forecasting result will not be very accurate because of the residuals. The next step is to repeat the estimate statements with diiferent non-seasonal autoregressive (p) and moving average (q) values.

Part 7

Autocorrelation Check of Residuals

To Lag	Chi-Square	DF	Pr > ChiSq	Autocorrelations					
6	69.81	2	<.0001	-0.210	0.003	0.300	-0.149	0.162	0.221
12	108.15	8	<.0001	-0.180	0.049	0.174	-0.234	0.050	0.052
18	164.05	14	<.0001	-0.254	0.149	-0.061	-0.244	0.160	-0.054
24	234.19	20	<.0001	-0.154	0.132	-0.161	-0.221	0.254	-0.192
30	295.09	26	<.0001	-0.171	0.209	-0.229	-0.037	0.133	-0.203
36	342.11	32	<.0001	0.032	0.147	-0.195	0.037	0.187	-0.206
42	403.89	38	<.0001	0.236	-0.015	-0.200	0.251	0.032	-0.142
48	457.41	44	<.0001	0.191	-0.067	-0.094	0.259	-0.162	-0.095

Figure 3-23 is shown in Part 8 of program1 and it displays residual correlation diagnostics for change of sales and Figure 3-24 is shown in Part 9 of program1 and it displays residual normality. Looking at ACF residuals, all of the spikes are not within significant limits, indicating that there is autocorrelation in the residuals; hence there is no white noise.

Part 8

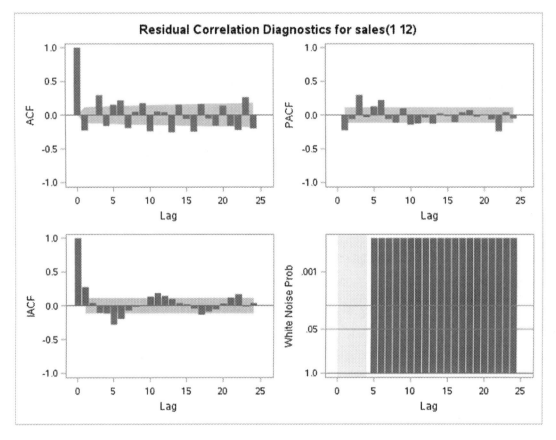

Figure 3-23. *Program1 Residual correlation for change of sales*

Part 9

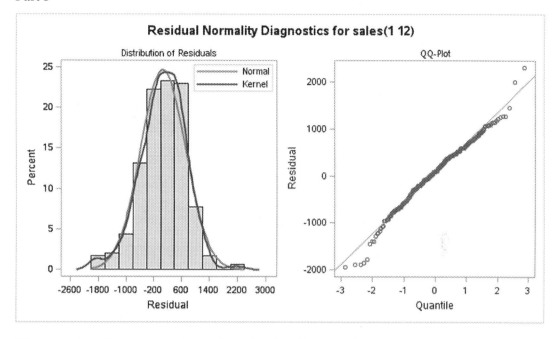

Figure 3-24. *Program1 Normality plot for change of sales*

Part 10 of program1 displays the forecast table for the next 30 months with observation number, forecast values, standard error, and lower and upper limits for a 95% of confidence interval for the forecast value.

Part 10

Forecasts for variable sales

Obs	Forecast	Std Error	95% Confidence Limits		Actual	Residual
310	60343.3298	649.7458	59069.8515	61616.8081	.	.
311	60620.7509	664.1855	59318.9712	61922.5307	.	.
312	65398.8061	690.0498	64046.3333	66751.2789	.	.
313	59613.7495	705.7882	58230.4300	60997.0690	.	.
314	56666.7033	721.9649	55251.6782	58081.7284	.	.
315	61227.4034	737.1826	59782.5519	62672.2548	.	.
316	60312.1223	752.1493	58837.9369	61786.3078	.	.

(*continued*)

Forecasts for variable sales

Obs	Forecast	Std Error	95% Confidence Limits		Actual	Residual
317	63023.1442	766.7800	61520.2830	64526.0054	.	.
318	61546.0300	781.1409	60015.0219	63077.0380	.	.
319	62866.1756	795.2393	61307.5352	64424.8161	.	.
320	62460.0913	809.0924	60874.2993	64045.8832	.	.
321	60811.1759	822.7120	59198.6900	62423.6618	.	.
322	62039.1728	906.7294	60262.0159	63816.3297	.	.
323	62323.7806	928.1567	60504.6269	64142.9342	.	.
324	66906.7211	952.9444	65038.9845	68774.4578	.	.
325	61499.9533	974.1240	59590.7053	63409.2012	.	.
326	58600.8408	995.1263	56650.4291	60551.2524	.	.
327	62913.4195	1015.4831	60923.1091	64903.7298	.	.
328	61948.0887	1035.4598	59918.6248	63977.5526	.	.
329	64788.9437	1055.0428	62721.0978	66856.7896	.	.
330	63305.3000	1074.2704	61199.7687	65410.8313	.	.
331	64697.2866	1093.1587	62554.7349	66839.8383	.	.
332	64247.8647	1111.7263	62068.9213	66426.8082	.	.
333	62500.4467	1129.9887	60285.7095	64715.1838	.	.
334	63802.5319	1178.8609	61492.0070	66113.0569	.	.
335	64091.7558	1201.2134	61737.4208	66446.0908	.	.
336	68631.5867	1225.1137	66230.4080	71032.7653	.	.
337	63316.9833	1247.0373	60872.8350	65761.1315	.	.
338	60432.0997	1268.7241	57945.4460	62918.7533	.	.
339	64689.0636	1289.9364	62160.8348	67217.2924	.	.

Figure 3-25 is shown in Part 11 of program1 and it displays the plot of the Forecast for sales for the next 30 months.

Part 11

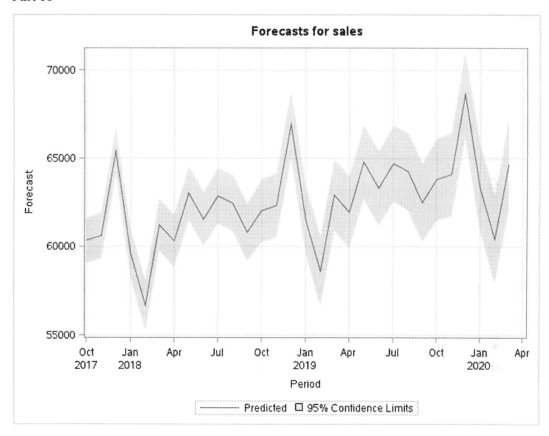

Figure 3-25. *Program1 Forecast sales plot of the retail sales data*

Let's try few other ARIMA models with different non-seasonal terms to identify the best models like ARIMA$(3,1,2)(2,1,2)_{12}$, ARIMA$(4,1,2)(2,1,2)_{12}$, ARIMA$(5,1,2)(2,1,2)_{12}$, ARIMA$(3,1,1)(2,1,2)_{12}$ and ARIMA$(6,1,1)(2,1,2)_{12}$ model. Out of all these models ARIMA$(6,1,1)(2,1,2)_{12}$ is the only model that has a lower AIC value as compared to other models and is able to pass the significance test; hence the ARIMA$(6,1,1)(2,1,2)_{12}$ model will be considered the final model for forecasting the retail sales.

```
/*Fitting seasonal ARIMA model for the retail sales with   different non
seasonal terms*/
/*by the maximum likelihood method*/
```

Program2:

```
    proc arima data = libref.sales;
    identify var = sales(1,12) stationarity= (adf) ;
    estimate p = (6)(12) q = (1)(12)  method = ml;

  /*Forecasting next 30 months*/

    forecast lead = 30 interval = month id = Period
    printall out = libref.pred_output;
    run;
```

In Program2 code **Identify** statement, **Stationarity = (adf)**. In the **estimation** of an ARMA model, the order of the autoregressive model with the non-seasonal and seasonal **p = (6) (12)** and the order of the moving average model with the non-seasonal and seasonal q = **(1) (12)** is mentioned. **Estimate** statement, **Method = ml**, **Forecast** statement, **lead = 30, interval = month, id = Period, out = libref.pred_output** are used for reasons explained previously for Program1.

The **Seasonal ARIMA output of program2** is spli into several sections and each of the is discussed below.

Part 1 of program2 displays the descriptive statistics for the retail sales series like period of differencing, mean of working series, standard deviation, number of observations, and number of observations eliminated by differencing.

Part 1

The ARIMA Procedure	
Name of Variable = sales	
Period(s) of Differencing	1,12
Mean of Working Series	5.222973
Standard Deviation	1050.084
Number of Observations	296
Observation(s) eliminated by differencing	13

Part 2 of program2 displays the table of the autocorrelation check for white noise and the autocorrelations are displayed in groups of 6, depending upon the number of lags; by default the number of lags is 24. To build an ARIMA model the values must be autocorrelated with its past values; if the values are not autocorrelated, then there is no point of building an ARIMA model. From the table it is clearly visible that all the p values in the respective lags are less than 0.05, indicating that there are autocorrelations; hence there is no white noise.

Part 2

Autocorrelation Check for White Noise

To Lag	Chi-Square	DF	Pr > ChiSq	Autocorrelations					
6	160.50	6	<.0001	-0.599	0.029	0.254	-0.270	0.099	0.167
12	235.79	12	<.0001	-0.261	0.083	0.171	-0.261	0.225	-0.148
18	261.18	18	<.0001	-0.058	0.149	-0.017	-0.165	0.166	-0.018
24	337.27	24	<.0001	-0.137	0.196	-0.092	-0.165	0.323	-0.197

Part 3 of program2 displays an Augmented Dickey-Fuller Tests table indicating that the data is stationary as all p values are less than 0.05.

Part 3

Augmented Dickey-Fuller Unit Root Tests

Type	Lags	Rho	Pr < Rho	Tau	Pr < Tau	F	Pr > F
Zero Mean	0	-471.802	0.0001	-34.34	<.0001		
	1	-1477.16	0.0001	-27.06	<.0001		
	2	-2523.29	0.0001	-15.28	<.0001		
Single Mean	0	-471.818	0.0001	-34.28	<.0001	587.63	0.0010
	1	-1477.66	0.0001	-27.02	<.0001	364.99	0.0010
	2	-2529.37	0.0001	-15.26	<.0001	116.41	0.0010
Trend	0	-471.817	0.0001	-34.22	<.0001	585.65	0.0010
	1	-1477.95	0.0001	-26.97	<.0001	363.76	0.0010
	2	-2533.02	0.0001	-15.23	<.0001	116.01	0.0010

Figure 3-26 is shown in Part 4 of program2 and it displays the stationary time series plot, autocorrelation function plot (ACF), partial autocorrelation function plot (PACF), and inverse autocorrelation function plot (IACF). These plots are used for trends and its autocorrelation analysis. Looking at these plots, whether the series is stationary or non-stationary can be identified. From the ACF plot it is clearly visible that the ACF plot decays very quickly, indicating that the series is stationary.

Part 4

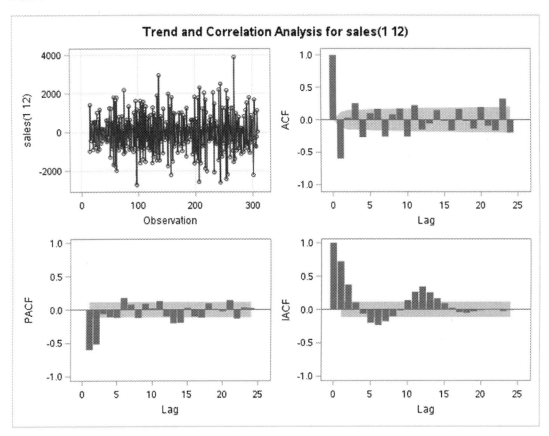

Figure 3-26. *Program2 stationary time series*

Part 5 of program2 displays results of the maximum likelihood estimation table. The mean term is represented as MU and its estimated value is 3.87909. The moving average term is represented as MA1,1 estimated value is 0.79134 and MA2,1 estimated value is 0.85889. The autoregressive term is represented as AR1,1 estimated value is 0.23121 and AR2,1 estimated value is 0.19106. The t values and p values are considered as the significance tests for parameter estimates and helps in identifying the insignificant and unnecessary terms in the model. In this case (MA1,1), (MA2,1), AR1,1, (AR2,1) term are considered as significant as t values are >2 and p values are less than 0.05 whereas MU does not add any meaning.

Part 5

Maximum Likelihood Estimation

Parameter	Estimate	Standard Error	t Value	Approx Pr > ltl	Lag
MU	3.87909	2.34894	1.65	0.0987	0
MA1,1	0.79134	0.03761	21.04	<.0001	1
MA2,1	0.85889	0.05471	15.70	<.0001	12
AR1,1	0.23121	0.06243	3.70	0.0002	6
AR2,1	0.19106	0.08003	2.39	0.0170	12

Part 6 of program2 displays a goodness-of-fit statistics table, which is used in comparing the models and selecting the best model. The Constant estimate is the function of the mean term MU, moving average and autoregressive term. The Variance estimate is the residual series variance and Std error estimate is the square root of the variance. AIC and SBC statistics are used to compare the models, and lower value of AIC and SBC indicate the better fitting model. In this Model AIC value is 4679.648 and SBC value is 4698.1, which is lower than the previous ARIMA model.

Part 6

Constant Estimate	2.41244
Variance Estimate	404718.9
Std Error Estimate	636.1752
AIC	4679.648
SBC	4698.1
Number of Residuals	296

Part 7 of program2 displays the correlations of parameters estimates table. In case two parameters are highly correlated with each other, one parameter must be dropped as it might impact the results. In this case none of the parameters are highly correlated with each other.

Part 7

Correlations of Parameter Estimates

Parameter	MU	MA1,1	MA2,1	AR1,1	AR2,1
MU	1.000	-0.021	0.044	-0.002	0.030
MA1,1	-0.021	1.000	-0.008	0.333	-0.041
MA2,1	0.044	-0.008	1.000	-0.139	0.647
AR1,1	-0.002	0.333	-0.139	1.000	-0.250
AR2,1	0.030	-0.041	0.647	-0.250	1.000

Part 8 of program2 displays an autocorrelation check of a residuals table.

Part 8

Autocorrelation Check of Residuals

To Lag	Chi-Square	DF	Pr > ChiSq	Autocorrelations					
6	51.68	2	<.0001	-0.227	0.097	0.222	-0.106	0.224	-0.006
12	66.52	8	<.0001	-0.059	-0.000	0.130	-0.163	-0.022	0.031
18	97.67	14	<.0001	-0.218	0.114	-0.074	-0.155	0.092	-0.032
24	136.59	20	<.0001	-0.077	0.051	-0.098	-0.192	0.196	-0.166
30	167.08	26	<.0001	-0.152	0.152	-0.180	0.005	0.027	-0.118
36	194.66	32	<.0001	0.007	0.107	-0.102	-0.012	0.180	-0.166
42	232.28	38	<.0001	0.214	-0.032	-0.136	0.200	0.020	-0.065
48	256.13	44	<.0001	0.093	-0.022	-0.049	0.180	-0.129	-0.083

Figure 3-27 is shown in Part 9 of program2. and it displays residual correlation diagnostics for change of sales; and Figure 3-28 is shown in Part 10 of program2 and it displays the residual normality.

Part 9

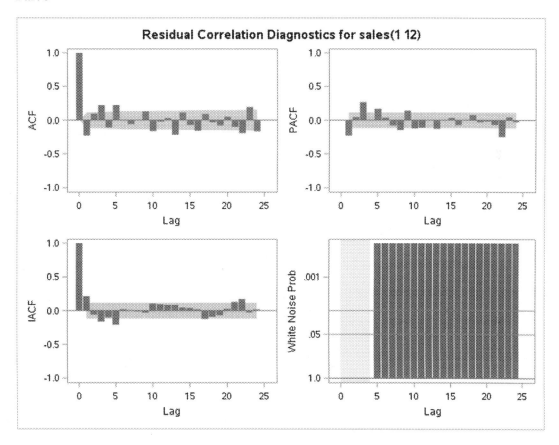

Figure 3-27. *Program2 Residual correlation for change of sales*

Part 10

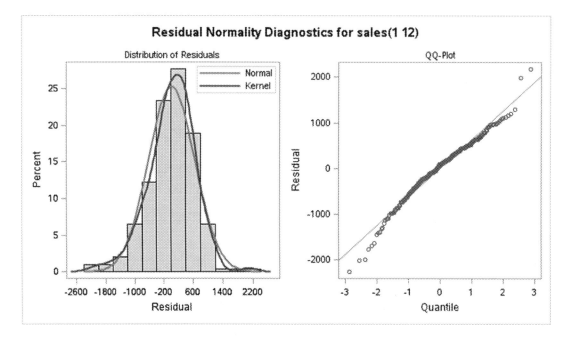

Figure 3-28. *Program2 Normality plot for change of sales*

Part 11 of program2 displays the forecast table for the next 30 months with observation number, forecast values, standard error, and lower and upper limits for a 95% of confidence interval for the forecast value.

Part 11

Forecasts for variable sales

Obs	Forecast	Std Error	95% Confidence Limits		Actual	Residual
310	60567.6037	636.1752	59320.7232	61814.4843	.	.
311	60611.7619	649.8775	59338.0255	61885.4984	.	.
312	65381.9718	663.2968	64081.9341	66682.0096	.	.
313	59559.8032	676.4499	58233.9859	60885.6206	.	.
314	56685.3684	689.3521	55334.2632	58036.4736	.	.
315	61382.6709	702.0172	60006.7426	62758.5993	.	.
316	60360.6722	755.7352	58879.4583	61841.8860	.	.

(continued)

Forecasts for variable sales

Obs	Forecast	Std Error	95% Confidence Limits		Actual	Residual
317	63091.0507	773.2064	61575.5940	64606.5074	.	.
318	61581.3891	790.2914	60032.4464	63130.3318	.	.
319	62902.5661	807.0148	61320.8461	64484.2861	.	.
320	62506.8816	823.3986	60893.0499	64120.7133	.	.
321	60843.3750	839.4628	59198.0583	62488.6918	.	.
322	62179.5163	933.6939	60349.5100	64009.5227	.	.
323	62371.2101	958.0449	60493.4766	64248.9436	.	.
324	66972.7441	981.7922	65048.4669	68897.0214	.	.
325	61529.5087	1004.9784	59559.7871	63499.2302	.	.
326	58612.6865	1027.6417	56598.5457	60626.8272	.	.
327	63030.9412	1049.8158	60973.3400	65088.5425	.	.
328	62000.2036	1084.3173	59874.9808	64125.4264	.	.
329	64876.8151	1107.7140	62705.7355	67047.8947	.	.
330	63359.8331	1130.6267	61143.8454	65575.8207	.	.
331	64757.9235	1153.0842	62497.9200	67017.9270	.	.
332	64308.1737	1175.1126	62004.9954	66611.3520	.	.
333	62538.5711	1196.7355	60193.0126	64884.1297	.	.
334	63925.2185	1252.5453	61470.2748	66380.1623	.	.
335	64156.7011	1278.5105	61650.8667	66662.5356	.	.
336	68729.9379	1303.9587	66174.2259	71285.6499	.	.
337	63366.3452	1328.9196	60761.7106	65970.9798	.	.
338	60439.5392	1353.4203	57786.8841	63092.1942	.	.
339	64801.9975	1377.4853	62102.1759	67501.8190	.	.

Figure 3-29 is shown in Part 12 of program2 and it displays the plot of the Forecast for sales for the next 30 months

Part 12

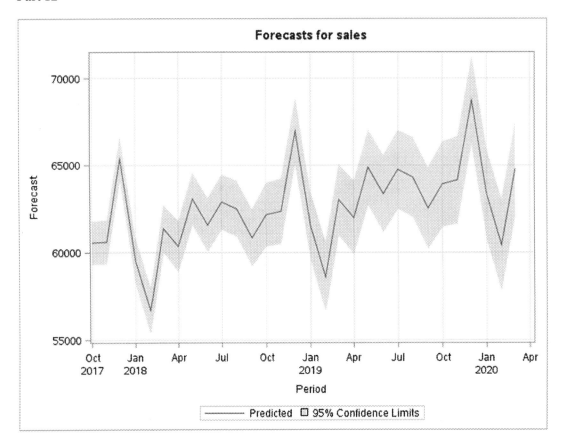

Figure 3-29. *Program2 Forecast sales plot of the retail sales data*

Summary

In this chapter we learned about the mathematical design of the ARIMA model and described the associated theoretical properties. We also presented a case study on sales forecasting by using a seasonal autoregressive integrated moving average (SARIMA) model. In this chapter it is clearly explained how SARIMA models are fitted in practice by following the identification, estimation, and diagnostic check stages of model construction in R and SAS.

References

1. Dunne, P.; Lusch, R., *Retailing*. Cengage Learning: 2007.

2. Bureau, U. S. C., Monthly and annual retail trade. Oct 02 2017 ed.;
    `https://www.census.gov/retail/index.html`.

3. Gensler, L. The world's largest retails 2017: Amazon & Alibaba
    closing in on Wal-Mart. `https://www.forbes.com/sites/`
    `laurengensler/2017/05/24/the-worlds-largest-retailers-`
    `2017-walmart-cvs-amazon/#63ba3b2320b5` (accessed Nov 04,
    2017).

4. Douglas, M. L.; Martha, C. C.; Janus, D. P., Supply Chain
    Management: Implementation Issues and Research
    Opportunities. *The International Journal of Logistics Management*
    1998, *9* (2), 1–20.

5. Levy, M.; Weitz, B. A.; Grewal, D., *Retailing Management*. 9 ed.;
    McGraw-Hill Companies, The: 2013; p 704.

6. Research, E. *Future of Retail Analytics*; `https://www.sas.com/`
    `content/dam/SAS/en_us/doc/research2/ekn-report-future-`
    `retail-analytics-106717.pdf`, 2013; p 40.

7. van Doorn, J.; Lemon, K. N.; Mittal, V.; Nass, S.; Pick, D.; Pirner,
    P.; Verhoef, P. C., Customer Engagement Behavior: Theoretical
    Foundations and Research Directions. *Journal of Service Research*
    2010, *13* (3), 253–266.

8. Tan, P. N.; Steinbach, M.; Kumar, V., *Introduction to Data Mining*.
    Pearson Education, Limited: 2014.

9. Jain, A. K., Data clustering: 50 years beyond K-means. *Pattern
    Recognition Letters* **2010**, *31* (8), 651–666.

10. Liu, V.; Curran, J. R. In *Web Text Corpus for Natural Language
    Processing*, European Chapter of association for computational
    linguistics (EACL), Trento, Italy, Trento, Italy, 2006.

11. Miranda, M. D.; Sassi, R. J., Using Sentiment Analysis to Assess Customer Satisfaction in an Online Job Search Company. In *Business Information Systems Workshops: BIS 2014 International Workshops, Larnaca, Cyprus, May 22-23, 2014, Revised Papers*, Abramowicz, W.; Kokkinaki, A., Eds. Springer International Publishing: Cham, 2014; pp 17–27.

12. Mohsen, A., RFID: an enabler of supply chain operations. *Supply Chain Management: An International Journal* **2007**, *12* (4), 249–257.

13. Angerer, A. The Impact of Automatic Store Replenishment Systems on Retail. University of St. Gallen, Austria, 2005.

14. Crone, T. P. K. a. S. F. In *Demand models for the static retail price optimization problem - A Revenue Management perspective*, OpenAccess Series in Informatics (OASIcs), Granado, P. C. D.; Joyce-Moniz, M.; Ravizza, S., Eds. Schloss Dagstuhl--Leibniz-Zentrum fuer Informatik: 2014; pp 101–125.

15. Gallego, G.; Wang, R., Multiproduct Price Optimization and Competition Under the Nested Logit Model with Product-Differentiated Price Sensitivities. *Operations Research* **2014**, *62* (2), 450–461.

16. Majeed, H.; Amir, Y., A new optimization model for market basket analysis with allocation considerations: A genetic algorithm solution approach. *Management & Marketing* **2017**, *12* (1), 1–11.

17. De Gooijer, J. G.; Hyndman, R. J., 25 years of time series forecasting. *International Journal of Forecasting* **2006**, *22* (3), 443–473.

18. Gerbing, D. Time Series Components. (accessed Nov 15, 2017).

19. Institute, S., *SAS/ETS 12.1 User's Guide*. Sas Inst: 2012.

20. Hyndman, R. J.; Athanasopoulos, G., *Forecasting: principles and practice*. OTexts: 2014.

CHAPTER 4

Telecommunication Case Study

The term telecommunication was first coined by Edouard Estaunie´ in 1904.[1] It is a fusion of Greek word "tele" meaning "distant" and the Latin word "communicatio," meaning "to share." Telecommunications is also known as telecom, and it is the science and technology of communication or sharing thoughts and ideas over a long distance. Since its inception in form of optical telegraph, modern-day telecommunication has evolved significantly both in terms of technology as well as an industry. Modern-day's telecommunication sector is a vast industry that comprises companies that provide hardware, software, and services necessary to facilitate information transfer.

Traditional telecommunications was basically about the Plain Old Telephony Services (POTS),which was only limited to voice communication and the only source of revenue generation for the industry, but with the help of technology, modern telecommunications is less about voice and more about messaging; email; digital information like music, video streaming, etc.[2] Value-added services play a critical role in the fast growth of the modern telecom sector. The Wireless industry is growing rapidly and is considered the biggest threat for the traditional telecom industry. The world's top telecommunications companies are China Mobile Ltd, Verizon Communications Inc, AT&T Inc, Vodafone Group plc, Nippon Telegraph & Telephone Corporation, Softbank Group Corp, Deutsche Telekom AG, Telefonica S.A, America Movil, and China Telecom.[3] The telecommunications services plays important roles in the economic development of the countries. Our day-to-day activities are heavily dependent on effective telecommunications services, for example, tele and Internet banking, airline tickets booking, hotel bookings by travel agencies, etc., can conduct their daily operations effectively because of the telecommunications services.

© Deepti Gupta 2018
D. Gupta, *Applied Analytics through Case Studies Using SAS and R*,
https://doi.org/10.1007/978-1-4842-3525-6_4

Types of Telecommunications Networks

In today's world, a vast amount of data is transferred across various types of data networks. Each network type has its own characteristics, advantages, and limitations. Efficient and cost-effective information transfer requires knowledge of available networks and their capabilities. Different types of data networks that exist are explained in the following section.[4]

1. Local area network (LAN): Local area network is defined as the group of computers and related devices or assets that share a common wireless link to a server within in a limited area such as an office building, school, university, residence, etc.

2. Wide area network (WAN): Wide area network connects different smaller networks including the local area networks and metro area networks. Wide area networks exist over the large geographical area.

3. Metropolitan area network (MAN): In Metropolitan area networks, the users can communicate with the computer resources in a geographical area larger than that covered by a local area network but smaller than the wide local network.

4. Internet area network (IAN): Internet area network is a type of communication network where the geographic form for a network is completely removed as applications and communications services are virtualized. In an Internet area network, voice and data endpoints are connected within a cloud environment over IP.

5. Campus area network (CAN): Campus area network is a type of network where two or more local area network are interconnected within the limited geographic area, for example, college building, office building, etc. Campus area network is also known as corporate area network.

6. Virtual private network (VPN): In a virtual private network, a safe and encrypted connection over a less secure network like the Internet is created, where data travels through secure tunnels and vpn users must use usernames and passwords in order to access the vpn .VPN will allow the remote users to securely access their corporate applications and other resources.

In this chapter, a case study on predicting the probability of customer churn for a telecommunication industry is presented using R and SAS Studio. A brief overview of key applications of analytics that have reshaped the telecommunication industry are discussed followed by an in-depth discussion about the decision tree model, its features, advantages, and limitations. The case study walks you through the main steps for data exploration, model implementation, and interpretation of outputs in R. In the subsequent section model, implementation in SAS Studio is demonstrated. Finally, a quick summary of the chapter is provided.

Role of Analytics in the Telecommunications Industry

The telecom industry is one of the most progressive and challenging industries. The high volatility of this market is due to brisk changes in technology, especially in wireless sectors. The fast-changing needs of the target audience make it necessary for the telecommunications firms to quickly adapt to the modern technology as well as align their marketing and sales strategies accordingly. The telecom sector relies heavily on advance analytics for a wide variety of applications that include network optimization, fraud identification, price optimization, predicting customer churn, and enhancing customer experience. An overview of some of these key applications is further discussed in this section.

Predicting Customer Churn

Analytics plays a significant role in predicting the customer churn, which implies identifying customers that have a high likelihood to terminate their relationship with the service provider. When postpaid or prepaid customers switch from one service provider to other service provider for better services, that process is known as customer churn.[5] In today's fast-paced marketplace, customer churn is the biggest problem that all of the telecom industry are facing and the common reasons for customer churn are high price, poor connection, bad customer service, outdated technology, and mobile number portability, etc. Customer churn causes significant financial loss to the telecommunication firms. Losing old and existing customers will directly impact the revenue and the growth of the telecom firm and acquiring new customers is always more expensive than retaining the old and existing customer.

Historical data of customer usage like number of calls made, number of sms, total call minutes, average bill amount, call drops, number of customer service calls, and customer demographic data are analyzed in order to predict the likelihood of customer churn prior to its actual occurrence. For example, cases having a high number of customer service calls are potentially facing persistent unresolved issues. If these cases are not provided with effective and quick customer service, there is a high probability it might result in customer churn. Social media like Facebook, Twitter, etc., are the platforms where customers write their experiences and feedback about the services used by them, and by extracting and analyzing social media feedback and reviews and conducting sentiment analysis help the telecom firms in identifying the reasons for customer churn or change of service.

Advance predictive analytics and modeling techniques like a decision tree, random forest, logistic regression, etc., helps the telecom sector to predict which customers have a higher propensity to churn. This also helps telecom firms identify any systemic issues or problems being faced by that particular segment of customers. Customer churn can be mitigated by resolving the identified issues and/ or providing those customers with the customized promotional offers, discounts, and incentives to reduce the probability of churn and increase customer retention. Figure 4-1 displays the customer retention strategies.

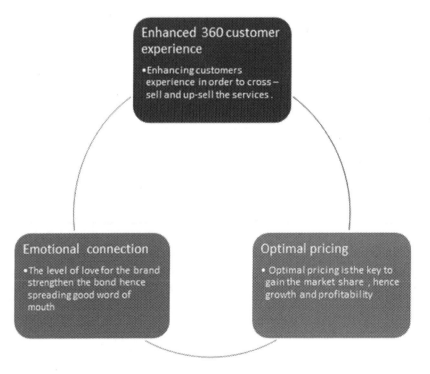

Figure 4-1. Customer retention strategies

To survive in highly competitive markets, almost every telecommunication company tracks consumers' past data usage and employs predictive analytics to predict the possible churners in advance and helps in mitigating customer churn to a large extent. Industry trends show that average annual churn in the telecom sector is between 20 to 40%.[6] By reducing the churn rate by as little as 1%, telecom companies can boost the revenue and achieve higher growth and profitability. A real-time example is Verizon wireless, which is the largest wireless service provider in the United States; they were able to reduce its customer churn by applying analytics and predictive models in their consumers' usage data.[7]

Network Analysis and Optimization

The next challenge for the telecommunication companies is to have healthy network and effective capacity planning to serve not only their existing customers but also cater for future demands.[8] Proper capacity planning and network optimization help customers to avail product and services at any point of time and any location without any network outage and issues. The objective of the telecommunication companies is to obtain the visibility of their networks that show how their internal and external customers are utilizing their network and how their network is performing. Proactive network analysis helps the telecommunication companies to manage the network without any congestion. Healthy networks make the customers happy and satisfied; poor network or any outage in network results in frequent call drops and poor sound quality, which will result in high attrition rates, making the customers unhappy and unsatisfied and spread the negative word of mouth for the service providers.

Real-time and predictive analytics enable telecommunications companies to analyze their consumer's behavior, which will help in creating the network usage polices for individual customers and identifying the new product offerings for them. Capacity planning is a technique applied to quantify the amount of bandwidth required to satisfy demand during peak hours.[9] Applying predictive analytics and modeling on historical network data help the operators to identify which part of the network is highly utilized in different times at different locations, which will help the telecom operators to make the decisions for adding the capacity at the right location at the right time and monitor the data traffic in a better way. Telecommunication companies can optimize their resource allocation toward different networks based on these models and ensure the network doesn't remain underused or overutilized.

Fraud Detection and Prevention

In the telecommunications industry the most fraudulent activities pertain to the usage of the carrier services without having the intention to pay for the service being used. Telecommunication fraud is considered as one of the biggest causes of loss of revenue in the telecom sector.[10] Fraud not only results in monetary loss, but it also impacts the brand value. Massive numbers of transactions and vast networks offer the opportunity to fraudsters to operate without detection.

The other challenge and the objective for the telecommunications industry is identifying the fraudsters responsible for committing different types of fraud.[11] Frauds can be directed directly toward monetary gains or could be directed toward gaining personal or financial information of the customers. The latter can be classified under identity theft. The telecommunications industry is comprised of different services like telephones, VOIP, Internet, etc.; therefore. fraud in this area is a broad topic. The most important types of telecom fraud include subscription fraud, internal fraud, sim card cloning fraud, and technical fraud.

Subscription fraud occurs when the client uses the telephone and Internet services without having any intention of paying; and this is one of the most frequently occurring frauds in telecommunication companies worldwide. Internal fraud occurs when the employee uses the internal and secure information about the system for their personal benefits.[12] Sim card cloning fraud involves copying the information inside the sim card of the client to the other empty sim card so the information can be used and can get access to the carrier services for their business as well as personal use and technical fraud occurs when the fraudsters for their personal benefits, utilizing the technical weakness of the system.[13] Predictive analytics and modeling techniques like decision tree, Bayesian network, neural network, and fuzzy rules are used in conducting the fraud analysis in order to detect the fraud in early stages and prevent the fraud happening in the telecommunications companies.[14] Effective fraud management strategies help telecommunications companies in maintaining their good brand image and save the companies from huge financial loss, hence driving the profitability and growth of the companies.

Price Optimization

Price optimization in a dynamic industry like telecommunications is critical for the success of business. Price optimization is the strategy by which the telecommunications companies can use the customer usage, network data, and demographic data to analyze the customers' behaviors and how sensitive customers are to the change in the price of the services offered to them and derive how much business can be achieved within defined profitability levels.[15] Optimal pricing plays a critical role in defining the profitability and growth of the business in telecommunications companies. Setting the right price for the services for the right customers at the right time is very important. If prices for the services are too low, it will result in the loss of revenue from the customer segment, which is less sensitive to the price and focus on the quality of the services, while setting the price too high can lead to reduced demand, which will also result in revenue loss. Predictive analytics and optimization techniques can help strike an optimal balance between price and volume. Figure 4-2 displays the price setting analytics element.

Figure 4-2. *Price setting analytics element*

The telecommunications industry is a highly competitive market so pricing optimization is not a one-time strategy but rather a continuous process. It is a critical part of the revenue optimization strategy. Telecom firms rely on price optimization to gain market share during the early phase of product or service in its life cycle. As the offering progresses through different stages in its life cycle, pricing is constantly optimized to maximize revenue and yet stay competitive. Unoptimized pricing plans might result in customers paying more for the services, leading to discontent and increasing their likelihood to churn.

Predictive analytics can also help in optimizing bundled offerings for customers and in turn promote cross-selling and up-selling. Bundle-based pricing, promotional-based pricing, usage-based pricing, etc., are a few of the strategies that all of the telecommunications companies are implementing to increase customer satisfaction as well as increasing cross-selling and up-selling opportunities.[16]

Case Study: Predicting Customer Churn with Decision Tree Model

A Decision tree is a type of supervised learning where there is a dependent or target variable and independent or explanatory variable. It is a graphical display of possible outcomes or solutions to a decision based on certain conditions. A decision tree is a tree-like diagram displayed upside down, which starts splitting with a single variable and then branches off into a number of solutions.[17] A decision tree can be used for both categorical and continuous dependent and independent variables. A decision tree is divided into two types based on the type of dependent variable; when a dependent variable is continuous, then it is called a continuous variable decision tree; and when a dependent variable is categorical, then it is called a categorical variable decision tree. There are several decision tree classification algorithms like ID3 algorithm, C4.5 algorithm, CART algorithm, SLIQ algorithm, etc. An ID3 algorithm can handle only categorical values whereas CART, C4.5, and SLIQ algorithms can handle both categorical and continuous values. Based on these algorithms, the classification rules are derived

A decision tree is comprised of three main components.[18]

1. Root Node: The top node of the tree is Root node and the tree starts splitting from this node; it indicates the best independent variable, which helps in predicting the dependent variable.

2. Internal Node: In Internal Node the independent variables are tested and each branch denotes an outcome of the test.

3. Leaf Node: The bottom part of the tree is Leaf Node. The Final Decision or outcome of the classification is taken in Leaf Node or Terminal Node and the outcome is represented as Yes or No.

Advantages and Limitations of the Decision Tree

Like all models, the decision tree has its advantages as well as certain limitations. The key advantages of the decision tree are listed in this section.

1. The visual aspect of a decision tree makes it easy to understand and interpret.

2. Decision tree is not influenced by missing values and outliers, so less time is spent in the data cleaning process.

3. Decision tree can be used in both types of data – categorical and numerical variables.

4. Decision tree is considered to be a non-parametric method so the assumption of a linear relationship between variables is not needed; it can be used in the data where there is a nonlinear relationship between the variables.

The decision tree also has some limitations that are mentioned in this section.

1. In a decision tree, there is the problem of overfitting, which impacts the model results and can be handled by applying limitations on the model parameter and pruning processes.

2. Decision tree is not adequate in applying regression and predicting the values of the continuous attribute.

3. Decision tree is more efficient when a correlation exists between the variables; it is not the best approach if the variables in the data are not correlated.

4. In decision tree there is a likelihood of duplication with the same subtree on diverse paths.

Handling Missing Values in the Decision Tree

Missing values in data are always challenging. They decrease the predictive accuracy of the model. Missing values can be present in both a training data set, which is used for building the model; and a testing data set, which is used for testing the model. There are various methods that are used to handle the missing values in a training set: Null Value Strategy, Prediction Model, C4.5 Strategy, Imputation with k-Nearest Neighbor, etc.[19] Methods that are used in handling missing values at prediction time or testing are Discard Testing Instance, Predictive Value Imputation, Distribution Based-Imputation and Reduced-Feature Models, Acquire Missing Values, etc.[20]

Handling Model Overfitting in Decision Tree

In a decision tree, overfitting is the major problem and the overfitting problem occurs when the model fits training data very well but does not perform well on validation data, which impacts the model accuracy results. A model is considered to be a good classification model when it not only fits the training data well, but it must also classify records accurately on unseen data. The overfitting problem can be solved by implementing prepruning and postpruning processes.[21] Pruning reduces the tree size by cutting the branches of the tree that has little influence to classify instances. Pruned trees are less complex and more accurate than the original decision tree.

Prepruning

In the prepruning process, the tree growth is terminated before generating a fully grown tree by applying an early stopping rule like stopping a leaf node from increasing when an observed impurity gain measure falls below a certain threshold, but choosing the right threshold value is not easy; if the threshold value is very high, it will result in an underfitted model; and a very low threshold value will result in an overfitted model. The advantage of the prepruning process is that it will prevent generation of very complex subtrees that will overfit the training data.

Postpruning

In the postpruning process, the tree is allowed to grow to its full size. The fully grown tree is trimmed in a bottom-up manner by replacing a subtree with a new leaf node or by using the most frequently used branch of subtree. The pruning of tree will stop when no further improvements are seen or observed. There are various pruning methods in decision trees like reduced error pruning, minimum error pruning, error-based pruning, cost complexity method,[22] and many more, but one of the most popular methods of postpruning is the cost complexity method, which is measured based on the two parameters – the size of the tree and error rate of the tree. Postpruning gives better results as compared to prepruning because in postpruning the tree is allowed to grow to full size and then a pruning decision is made based on a full-grown tree; while in the prepruning process, early termination of the tree impacts the pruning decisions; hence the postpruning process is advisable in a decision tree model.

How the Decision Tree Works

To demonstrate how a decision tree works, let's consider a sample of 1,068 customers with seven variables: Number_customer_service_calls, Gender, International_plan, Voice_mail_plan, Total_night_minutes, Total_morning_minutes, and Churn. Churn is a dependent variable and all of the other six variables are independent variables. Out of 1,068 customers there is a 5.24% customer churn. Now a model will be created to predict which customers will churn based on their usage.

In this example we will segregate customer churn based on highly significant independent variables among all six variables. In such a scenario, the decision tree model is used. The first split is based on the Number_customer_service_calls, which signify that Number_customer_service_calls is a highly significant independent variable. Figure 4-3 displays the pruned decision tree model, which provides more detail about the nodes and split of the decision tree.

There are 1,068 customers for whom Number_customer_service_calls < 3.040 (node 1), and the customer churn is around 5.24%. The 1,068 customers for whom Number_customer_service_calls < 3.040 are further subdivided based on the International_plan variable. Out of 101 there are 33.66% of the customer churn for whom International_plan = yes. The 1,068 customers for whom International_plan = yes are further subdivided on the basis of Total_night_minutes variable. Out of 65 customers there is a 46.15% customers churn whose International_plan = yes and Total_night_minutes is < 216.898. Contrast to these customers whose International_plan = No, there is a 2.28% customer churn. In summary, based on the decision tree splits, customers who have a high likelihood of churn are those whose International_plan = yes and Total_night_minutes is < 216.898.

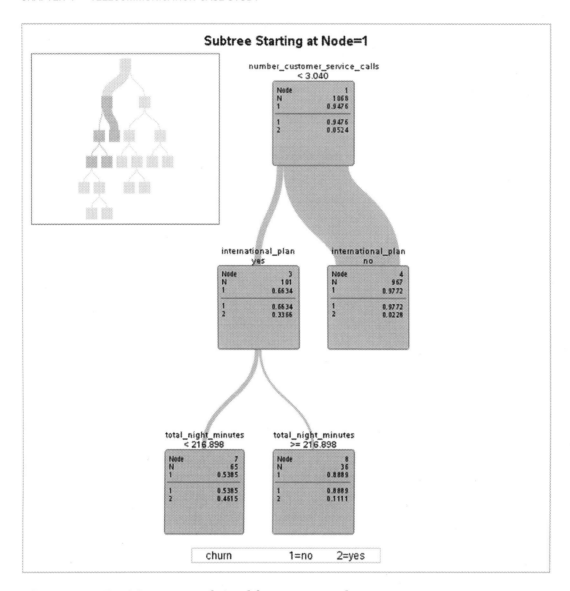

Figure 4-3. Decision tree explained for customer churn

Measures of Choosing the Best Split Criteria in Decision Tree

The key in a decision tree is how to select the best split as the accuracy of the tree is heavily depended on the split. In Figure 4-3 we have seen that the tree starts splitting at Number_customer_service_calls, hence Number_customer_service_calls is the highly significant variable. Now the question is how the tree recognizes the variable and the best split. There are various algorithms that are used in deciding splitting criterion, and a few of the algorithms are discussed below. Algorithms differ based on the type of dependent variable; as for the categorical variable there are different split algorithms like Gini Index, Entropy, and Information Gain; and for a continuous dependent variable, the Variance Reduction Split method is used.[23]

1. **Gini Index:** Gini Index is an impurity measure used in building a decision tree in a CART algorithm.[24] The CART algorithm will build a decision tree when a dependent variable is binary, which means it has only two values (yes or no). A variable having the higher GINI gain will be the one where the node will start splitting. The formula of a GINI gain of a split is provided below:

$$\text{GINI}(s,t) = \text{GINI}(t) - P_L\,\text{GINI}(t_L) - P_R\,\text{GINI}(t_R)$$

Where

s = split

t = node

$\text{GINI}(t)$ = Gini Index of input node t

P_L = Proportion of observation in Left Node after split, s

$\text{GINI}(t_L)$ = Gini of Left Node after split, s

P_R = Proportion of observation in Right Node after split, s

$\text{GINI}(t_R)$ = Gini of Right Node after split, s

In the above example, a decision tree was created based on the dependent variable Churn where customers churn (Yes) and which customers do not churn (No); so referring the same example we will split the customers using two independent variables International_plan and Number_customer_service_calls. Let's take the subset of the customers to make our calculation easy to understand. Out of 1,068 customers, let's study only 60 customers who are having customer churn and who are not having customer churn based on two independent variables International_plan and Number_customer_service_calls. Figure 4-4 displays the split based on International_plan and Number_customer_service_calls

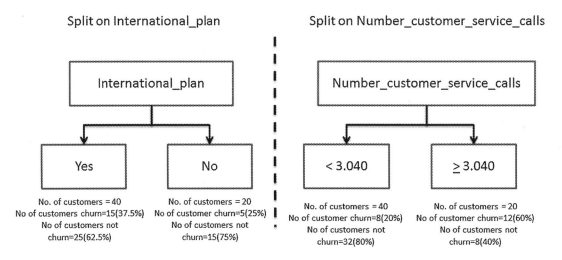

No. of customers = 60
No. of customers churn =20
No. of customers not churn =40

Split on International_plan   Split on Number_customer_service_calls

International_plan   Number_customer_service_calls

Yes   No   < 3.040   ≥ 3.040

No. of customers = 40   No. of customers = 20   No. of customers = 40   No. of customers = 20
No of customers churn=15(37.5%) No of customer churn=5(25%)   No of customer churn=8(20%)   No of customer churn=12(60%)
No of customers not   No of customers not   No of customers not   No of customers not
churn=25(62.5%)   churn=15(75%)   churn=32(80%)   churn=8(40%)

Figure 4-4. *Split on International_plan & Number_customer_service_calls*

Calculation of split on International_plan:

1. Gini for node **International_Plan**

$$=1-\left(\frac{40}{60}\right)^2-\left(\frac{20}{60}\right)^2=\mathbf{0.444}$$

2. Gini for sub-node **International_plan = Yes**

$$=1-\left(\frac{25}{40}\right)^2-\left(\frac{15}{40}\right)^2=\mathbf{0.468}$$

3. Gini for sub-node **International_plan = No**

$$=1-\left(\frac{15}{20}\right)^2-\left(\frac{5}{20}\right)^2=\mathbf{0.375}$$

4. Gini gain for Split **International_plan**

$$=0.444 - (40/60)*0.468 - (20/60)*0.375 = \mathbf{0.007}$$

Calculation of split on Number_customer_service_calls:

1. Gini for node **Number_customer_service_calls**

$$= 1 - \left(\frac{40}{60}\right)^2 - \left(\frac{20}{60}\right)^2 = \mathbf{0.444}$$

2. Gini for sub-node **Number_customer_service_calls < 3.040**

$$= 1 - \left(\frac{32}{40}\right)^2 - \left(\frac{8}{40}\right)^2 = \mathbf{0.320}$$

3. Gini for sub-node **Number_customer_service_calls >= 3.040**

$$= 1 - \left(\frac{8}{20}\right)^2 - \left(\frac{12}{20}\right)^2 = \mathbf{0.480}$$

4. Gini gain for Split **Number_customer_service_calls**

$$= 0.444 - (40/60)*0.320 - (20/60)*0.480 = \mathbf{0.071}$$

From the above calculation the Gini gain for split on **Number_customer_service_calls (0.071)** is higher than the split on **International_plan (0.007)**, therefore **Number_customer_service_calls** is considered the most significant variable and the decision tree node split will start from **Number_customer_service_calls**.

2. **Entropy**: Entropy is a measure of disorder used in building a decision tree in ID3 and C4.5 algorithms.[25] ID3 and C4.5 algorithms are applied on categorical dependent variables. ID3 uses binary splits whereas a C4.5 algorithm uses a multi-way split. The lower is the value of the entropy, the higher is the homogeneity; hence the variable having the lower entropy will be the one where the node will start splitting and will be considered as the best split in the decision tree. In other words, the less is the value of Entropy and the better is the classification. When a sample is homogeneous or all members of S belong to the same class, then the Entropy is equal to zero (pure node); when the collection is equally divided, it contains an equal number of

class p and class q than the Entropy and is equal to one (impure node); and when the collection has an unequal number of class p and class q, then the entropy is between 0 and 1 (impure node). Entropy is calculated as:

Entropy(S) $= -p \log_2 p - q \log_2 q$

   Where p = proportion of p class in S

         q = proportion of q class in S

         $\log_2$ = logarithm to the base 2

Entropy for **Parent node** = - (20/60) $\log_2$ (20/60) – (40/60) $\log_2$ (40/60) = **0.92**

Calculation of split on International_plan:

1. Entropy for sub-node **International_plan = Yes** = - (15/40) $\log_2$ (15/40) – (25/40) $\log_2$ (25/40) = **0.95**

2. Entropy for sub-node **International_plan = No** = - (5/20) $\log_2$ (5/20) – (15/20) $\log_2$ (15/20) = **0.81**

3. Weighted Entropy for Split **International_plan** = (40/60)*0.95 + (20/60)*0.81 = **0.90**

Calculation of split on Number_customer_service_calls:

1. Entropy for sub-node **Number_customer_service_calls < 3.040** = - (8/40) $\log_2$ (8/40) – (32/40) $\log_2$ (32/40) = **0.72**

2. Entropy for sub-node **Number_customer_service_calls >= 3.040** = - (12/20) $\log_2$ (12/20) – (8/20) $\log_2$ (8/20) =**0.97**

3. Weighted Entropy for Split **Number_customer_service_calls** = (40/60)*0.72 + (20/60)*0.97 = **0.80**

From the above calculation the Entropy for split on **Number_customer_service_calls (0.80)** is lower than the split on **International_plan (0.90),** therefore **Number_customer_service_calls** is considered the most significant variable and the decision tree node split will start from **Number_customer_service_calls**.

3. **Information gain**: Information gain is the measure of change in Entropy. The variable having the higher information gain value will be selected for the split. Information gain is defined as the differences of the Entropy for the parent node and Entropy for the child node. Information Gain is calculated as:

    Information Gain = *Entropy for parent node – Entropy for child node*

    Let's calculate the Information gain by considering the Entropy calculated for parent node and child node for **International_plan** and **Number_customer_service_calls**.

    Information Gain for **International_plan** = $0.92 - 0.90 = 0.02$

    Information Gain for **Number_customer_service_calls** = $0.92 - 0.80 = 0.12$

    From the above calculation the Information Gain for **Number_customer_service_calls (0.12)** is higher than the Information Gain for **International_plan (0.02),** therefore **Number_customer_service_calls** is considered the most significant variable and the decision tree node split will start from **Number_customer_service_calls**.

4. **Variance Reduction:** Variance Reduction method is used when the dependent variable is continuous, and it is considered as regression purity.[26] In the Variance Reduction method the standard formula of variance is used. The split having the lower variance will be considered as the best split and the decision tree will start splitting from that node. Variance Reduction method is calculated as:[27]

    Variance Reduction $= \dfrac{\Sigma (X - \bar{X})^2}{N}$

    Where X is real values

    X bar is mean of the values

    N is total number of observations

To calculate the variance, let's consider the number of customer churn as 1 and number of customer not churn as 0.

1. Mean of **Root Node**: There is a number of customers churn = 20 (1) and number of customers not churn = 40(0) so the Mean calculation it is written as (20*1+40*0)/60 = 0.33, and hence the variance is calculated as:

    Variance of **Root Node** = (20*(1-0.33) ^2+40*(0-0.33) ^2)/ 60 = **0.222**

2. Mean of sub-node **International_plan= Yes:** There is a number of customers churn = 15(1) and number of customers not churn = 25(0). so for the Mean calculation, it is written as (15*1+25*0)/40 = 0.375, and hence the variance is calculated as:

    Variance of sub-node **International_plan= Yes** = (15*(1-0.375) ^2+25*(0-0.375) ^2)/40 = **0.23**

3. Mean of sub-node **International_plan= No**: There is a number of customers churn = 5(1) and number of customers not churn = 15(0), so for the Mean calculation it is written as (5*1+15*0)/20 = 0.25, and hence the variance is calculated as:

    Variance of sub-node **International_plan= No** = (5*(1-0.25) ^2+15*(0-0.25) ^2)/20 = **0.18**

4. Weighted variance of sub-node **International_plan**

    = (40/60)*0.23+ (20/60)*0.18 = **0.213**

5. Mean of sub-node **Number_customer_service_calls < 3.040:** There is a number of customers churn = 8(1) and number of customers not churn = 32(0), so for the Mean calculation it is written as (8*1+32*0)/40 = 0.2, and hence the variance is calculated as:

    Variance of sub-node **Number_customer_service_calls < 3.040**

    = (8*(1-0.2) ^2+32*(0-0.2) ^2)/40 = **0.16**

6. Mean of sub-node **Number_customer_service_calls >= 3.040:** There is a number of customers churn = 12(1) and number of customers not churn = 8(0) so for the Mean calculation, it is written as $(12*1+8*0)/20 = 0.6$, and hence the variance is calculated as:

    Variance of sub-node **Number_customer_service_calls >= 3.040**

    $$= (12*(1-0.6)^2 + 8*(0-0.6)^2)/20 = \textbf{0.24}$$

7. Weighted variance of sub-node **Number_customer_service_calls**

    $$= (40/60)*0.16 + (20/60)*0.24 = \textbf{0.186}$$

From the above calculation the variance for **Number_customer_service_calls (0.186)** is lower than the variance for **Root Node (0.222)**; therefore **Number_customer_service_calls** is considered the most significant variable and the decision tree node split will start from **Number_customer_service_calls**.

Decision Tree Model Using R

Business Problem: To predict the probability of the customer churn.

   Business Solution: To build the Decision tree model.

About Data

In this telecommunications case study, in order to illustrate the likelihood of customer churn, the decision tree model is created and the data is generated synthetically. In this dataset there are a total of 1,000 observations and 14 variables: 3 variables are numeric and 11 variables are categorical.

   The churn_dataset contains information on 1,000 customers. Churn is the dependent or target variable in the data, where Yes denotes customer churn and No denotes customer does not churn. There are 74% of customers who do not churn and 26% of customer who do churn in this data. This data set is used to create the decision tree model in order to predict the probability of the customer churn.

In R, Create your own working directory to import the dataset.

```
#Read the data from the working directory, create your own working directly
to read the dataset.

setwd("C:/Users/Deep/Desktop/data")

data1 <- read.csv ("C:/Users/Deep/Desktop/data/
churn_dataset.csv",header=TRUE,sep=",")

data2<-data.frame(data1)
```

Performing Data Exploration

In Exploratory data analysis, we are going to take a broad look at patterns, trends, summary, outliers, missing values, and so on in the existing data. R codes for data exploration and their output is discussed in the section below.

```
#perform exploratory data analysis to know about the data

# display top 6 rows of dataset to see how data look like

head (data2)
```

	Sex	Marital_Status	Term	Phone_service	International_plan
1	Female	Married	16	Yes	Yes
2	Male	Married	70	Yes	No
3	Female	Married	36	Yes	No
4	Female	Married	72	Yes	No
5	Female	Married	40	Yes	Yes
6	Female	Single	15	Yes	Yes

	Voice_mail_plan	Multiple_line	Internet_service	Technical_support
1	Yes	No	Cable	Yes
2	Yes	No	Cable	Yes
3	Yes	No	Cable	Yes
4	No	Yes	Cable	Yes
5	No	Yes	Cable	No
6	Yes	No	No Internet	No internet

	Streaming_Videos	Agreement_period	Monthly_Charges	Total_Charges	Churn
1	No	Monthly contract	98.05	1410.25	Yes
2	Yes	One year contract	75.25	5023.00	No
3	Yes	Monthly contract	73.35	2379.10	No
4	Yes	One year contract	112.60	7882.25	No
5	Yes	Monthly contract	95.05	3646.80	No
6	No internet	Monthly contract	19.85	255.35	No

display bottom 6 rows to see how data look like

tail(data2)

	Sex	Marital_Status	Term	Phone_service	International_plan
995	Male	Single	59	No	No
996	Female	Married	56	Yes	No
997	Female	Married	56	Yes	No
998	Male	Married	57	Yes	No
999	Male	Married	51	Yes	No
1000	Male	Single	69	Yes	yes

	Voice_mail_plan	Multiple_line	Internet_service	Technical_support
995	No	No phone	DSL	Yes
996	Yes	Yes	Fiber optic	No
997	Yes	Yes	DSL	No
998	Yes	No	Fiber optic	No
999	Yes	No	DSL	Yes
1000	Yes	No	No Internet	No internet

	Streaming_Videos	Agreement_period	Monthly_Charges	Total_Charges	Churn
995	No	One year contract	44.45	2145.00	No
996	Yes	One year contract	84.50	4054.20	No
997	Yes	Two year contract	83.50	3958.25	No
998	No	One year contract	73.95	4326.80	No
999	No	One year contract	60.90	3582.40	No
1000	No internet	Two year contract	19.15	1363.45	No

describe the structure of data, it displays the datatype of each variable present in the data like whether that particular variable is numeric, factor etc.

```
str(data2)
```

```
'data.frame':       1000 obs. of  14 variables:
 $ Sex               : Factor w/ 2 levels "Female","Male": 1 2 1 1 1 1 2 1
                      2 1 ...
 $ Marital_Status    : Factor w/ 2 levels "Married","Single": 1 1 1 1 1 2 1
                      1 1 1 ...
 $ Term              : int   16 70 36 72 40 15 1 36 5 57 ...
 $ Phone_service     : Factor w/ 2 levels "No","Yes": 2 2 2 2 2 2 2 2 2 2 ...
 $ International_plan : Factor w/ 3 levels "No","yes","Yes": 3 1 1 1 3 3 3 3
                      3 1 ...
 $ Voice_mail_plan   : Factor w/ 2 levels "No","Yes": 2 2 2 1 1 2 2 1 2 1 ...
 $ Multiple_line     : Factor w/ 3 levels "No","No phone ",..: 1 1 1 3 3 1
                      3 3 3 3 ...
 $ Internet_service  : Factor w/ 4 levels "Cable","DSL",..: 1 1 1 1 1 4 1 1
                      1 1 ...
 $ Technical_support : Factor w/ 3 levels "No","No internet ",..: 3 3 3 3 1
                      2 1 1 1 3 ...
 $ Streaming_Videos  : Factor w/ 3 levels "No","No internet ",..: 1 3 3 3 3
                      2 1 3 3 3 ...
 $ Agreement_period  : Factor w/ 3 levels "Monthly contract",..: 1 2 1 2 1
                      1 1 1 1 2 ...
 $ Monthly_Charges   : num   98 75.2 73.3 112.6 95 ...
 $ Total_Charges     : num   1410 5023 2379 7882 3647 ...
 $ Churn             : Factor w/ 2 levels "No","Yes": 2 1 1 1 1 1 1 2 2 1 ...
```

```
#display the column name of the data
```

```
names(data2)
```

```
[1] "Sex" "Marital_Status"     "Term"               "Phone_service"
[5] "International_plan"  "Voice_mail_plan"    "Multiple_line"
[8] "Internet_service"   "Technical_support"  "Streaming_Videos"
[11]"Agreement_period"   "Monthly_Charges"    "Total_Charges"
[14]"Churn"
```

```
# display the datatype

class(data2)
[1] "data.frame"
#display the summary or descriptive statistics of the data

  summary(data2$Monthly_Charges)

  Min.   1st Qu  Median   Mean   3rd Qu    Max.
  18.95   40.16   74.72   66.64   90.88   116.25
```

Splitting Data Set into Training and Testing

In this section the data is split into two parts – train data set and test data set; the splitting ratio is 70:30; it means that 70% of data contributes to the train dataset and 30% of data contributes to the test dataset. The train dataset is used to build the model and test dataset is used to test the performance of the model.

```
#Set seed in order to reproduce the sample

set.seed(123)

#splitting data set into training and testing dataset in 70:30

sample.split() is the function of the package "caTools",hence the package
is installed below before splitting the dataset.

install.packages("caTools")

library(caTools)

sample <- sample.split(data2$Churn,SplitRatio=0.70)

#No of observations in train dataset

train_data <- subset(data2,sample==TRUE)

# No of observations  in test dataset

test_data <- subset(data2,sample==FALSE)
```

Churn_dataset is divided into two parts train_dataset and test_dataset, the splitting ratio is 70:30, it means 70% of data contributes to the train_dataset and 30% of data contributes to the test_dataset. The train_dataset is used to build the model and the test_dataset is used to test the performance of the model.

Model Building & Interpretation on Training and Testing Data

In the following program1 code **rpart package** is used to grow the decision tree.

```
#Growing full decision tree model by using rpart on training data
```

```
install.packages("rpart")
```

```
library(rpart)
```

Program1:

```
churn_model <- rpart(Churn ~ ., data=train_data,
method = "class", parms = list(split = 'information'),cp=-1)
```

```
churn_model
```

The decision tree model is created by using the rpart function with dependent variable Churn and all other independent variables in the data, method = class is used to set the method used and here it is class, parms is used in setting the parameter.

For splitting criterion as split =information by default in rpart, it is gini impurity, different splitting criterion, and an algorithm that is already discussed in detail.

The complexity parameter (cp) is specified in negative value as -1 to ensure that the decision tree is grown fully. Following are a few rules to be mentioned for a fully grown decision.

```
n= 700
node), split, n, loss, yval, (yprob)
      * denotes terminal node

1) root 700 181 No (0.741428571 0.258571429)
2) Agreement_period=One year contract,Two year contract 314
   18 No (0.942675159 0.057324841)
4) Monthly_Charges< 89.025 213 4 No (0.981220657 0.018779343)
8) Total_Charges>=897.875 166   1 No (0.993975904 0.006024096)
16)Internet_service=DSL,Fiber optic,No Internet 151   0 No
   (1.000000000 0.000000000) *
17)Internet_service=Cable 15 1 No (0.933333333 0.066666667) *
9) Total_Charges< 897.875 47   3 No (0.936170213 0.063829787)
18) Total_Charges< 769.775 40   1 No (0.975000000 0.025000000)
36) Term>=12.5 27   0 No (1.000000000 0.000000000) *
37) Term< 12.5 13   1 No (0.923076923 0.076923077) *
19) Total_Charges>=769.775 7 2 No (0.714285714 0.285714286) *
5) Monthly_Charges>=89.025 101 14 No (0.861386139 0.138613861)
10) Term>=62.5 56   3 No (0.946428571 0.053571429)
20) Term< 69 25   0 No (1.000000000 0.000000000) *
21) Term>=69 31   3 No (0.903225806 0.096774194)
42) Technical_support=No 9   0 No (1.000000000 0.000000000) *
43) Technical_support=Yes 22   3 No (0.863636364 0.136363636)
86) Term>=71.5 13   1 No (0.923076923 0.076923077) *
87) Term< 71.5 9   2 No (0.777777778 0.222222222) *
```

#Installing rattle package and helper package like rpart.plot and
RcolorBrewer package in order to create the decision tree

```
install.packages("rattle")
library(rattle)

install.packages("rpart.plot")
library(rpart.plot)

install.packages("RColorBrewer")
library(RColorBrewer)
```

185

```
#Creating the decision tree plot by using fancyRpartPlot function

fancyRpartPlot(churn_model)
```

FancyRpartPlot is used to plot the decision tree. A fully grown decision tree is displayed in Figure 4-5.

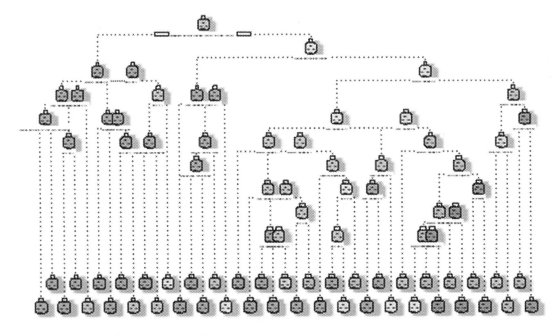

Figure 4-5. *Fully grown Decision tree*

In the following program1.1 code rpart function, Churn, method = class, parms, split = information is already discussed in the previous Program1 code.

```
#List of options that control the rpart algorithm and are ways preventing a
model from overfitting.
```

Program1.1

```
tree_model <- rpart(Churn ~ ., data=train_data,
method = "class",parms = list(split = 'information'),  maxdepth = 3,
minsplit = 2, minbucket = 2 )

tree_model
```

There are list of options that control the r part algorithm so the model overfitting can be prevented, and the overfitted model always impacts the predictive accuracy of the model. maxdepth is used to restrict the maximum depth of any node of the final tree, root node will be taken as depth 0, maxdepth = 3 is used to restrict the tree node at 3. Minsplit is used to set the minimum number of observations at a node for splitting to be feasible. Minbucket is used to set the minimum number of observations in any terminal or leaf node. In this case the tree is having two splits and two terminal or leaf nodes.

```
#Creating the decision tree plot
```

```
fancyRpartPlot(tree_model)
```

Figure 4-6 displays decision tree grown as depth 3. The decision tree nodes and sub-nodes explanation will be discussed in a later section.

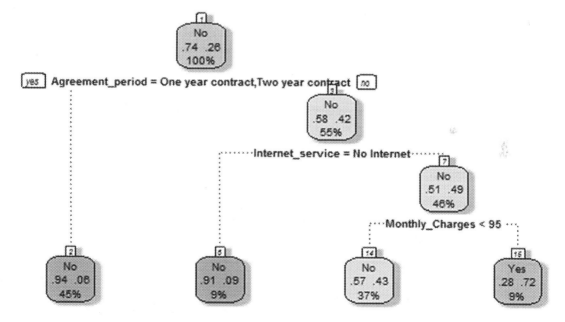

Figure 4-6. *Decision tree node as depth 3*

```
#Plotting it inorder to find cp value for pruning
```

```
  plotcp(tree_model)
```

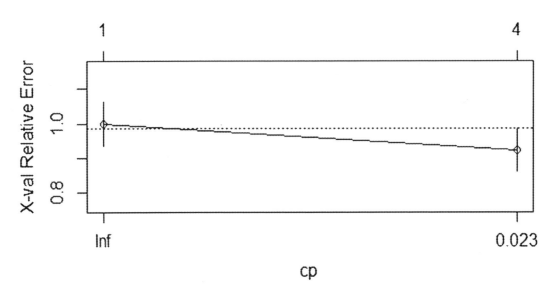

Figure 4-7. *Plot of size of tree vs. cp*

The fully grown tree will have overfitting issues, so to handle this issue the pruning process is performed; now the question is how to set the value for tree pruning so Figure 4-7 displays the plotcp() that will display the plot of size of tree vs, cp, and the least error point in plot will display the cp value.

```
#Finding the cross validation results

 printcp(tree_model)
```

```
Classification tree:
 rpart(formula = Churn ~ ., data = train_data,
 method = "class", parms = list(split = "information"),
 maxdepth = 3, minsplit = 2, minbucket = 2)
```

```
Variables actually used in tree construction:
[1] Agreement_period Internet_service Monthly_Charges
```

```
Root node error: 181/700 = 0.25857
```

```
n= 700
```

	CP	nsplit	rel error	xerror	xstd
1	0.051565	0	1.0000	1.00000	0.064002
2	0.010000	3	0.8453	0.92265	0.062301

printcp() will display the cross validation results. In this case study there are around 26% of training cases that are having customer churn. c, nsplit, rel error, cross validation error(xerror), and standard deviation (xstd) are displayed above. The complexity parameter (cp) of the smallest tree that is with the least error and standard deviation of the decision tree must be selected for pruning. In this case study the least xerror is 0.92265 with a standard deviation of 0.062301 and cp is 0.010000 so the tree is pruned by considering the cp value somewhat greater than 0.01

In program 1.2 code the fully grown tree will have overfitting issues so to handle this issue the pruning process is performed by using prune and by using cross validation results from printcp() function, complexity parameter (cp) of smallest tree that is with the least error and standard deviation of the decision tree is used.

Program 1.2

#Pruned model

prune_model <- prune(tree_model, cp=.02)

prune_model

In this case study the least xerror is **0.92265** with standard deviation of **0.062301** and cp is **0.010000** so the tree is pruned by considering the cp value somewhat greater than **0.01, cp = 0.2** is used to prune the tree for the best results . Below are a few rules mentioned for the pruned decision tree. The pruned decision tree always perform better than the original tree.

n= 700

node), split, n, loss, yval, (yprob)
      * denotes terminal node

```
1) root 700 181 No (0.74142857 0.25857143)
2) Agreement period=One year contract, Two year contract
   314 18 No (0.94267516 0.05732484) *
3) Agreement_period=Monthly contract 386 163 No
   (0.57772021 0.42227979)
6) Internet_service=No Internet64 6 No(0.90625000 0.09375000*
```

```
 7) Internet_service=Cable, DSL, Fiber optic 322 157 No
     (0.51242236 0.48757764)
14) Monthly_Charges< 95.125 258 111 No
     (0.56976744 0.4302325) *
15) Monthly_Charges>=95.125 64 18 Yes
     (0.28125000 0.71875000) *
```

To explain how a pruned decision tree displayed in **Figure1-8** works is based on the rules generated above. There are **700 observations and 14 variables** from training datasets that are considered. Churn is a dependent variable and all the other **13** variables are independent variables. Out of **700** customers there is a **26%** customers churn. Now a model will be created to predict which customers will churn based on their usage. In this example we will segregate customer churn based on a highly significant independent variable among all **13** variables. The first split is based on **Agreement_period**; it means **Agreement_period** is considered a highly significant variable in the model.

There are **700** customers from the training dataset where in root node **74%** of customers do not churn and **26%** customers churn. Customers for whom **Agreement_period= Monthly contract**, the customer churn is **42.22%.** The customers for whom **Agreement_period= Monthly contract** are further subdivided based on the **Internet_service** variable. There is **48.75%** of customer churn for whom **Internet_service=Cable, DSL, Fiber optic**. The customers for whom **Internet_service=Cable, DSL, Fiber optic** are further subdivided on the basis of the **Monthly_Charges** variable. There is a **71.87%** customer churn for whom **Internet_service=Cable, DSL, Fiber optic,** and **Monthly_Charges is >= 95.125**. Contrast to these customers for whom **Internet_service = No**, there is a **9.37%** customer churn. In summary, based on the decision tree splits, customers who have a high likelihood of churn are those for whom **Internet_service=Cable, DSL, Fiber optic, and Monthly_Charges is >= 95.125**.

```
#Creating the pruned decision tree plot
fancyRpartPlot(prune_model)
```

Figure 4-8 displays the pruned decision tree model and the explanation of the decision tree node and sub-node based on the rules generated in the section above.

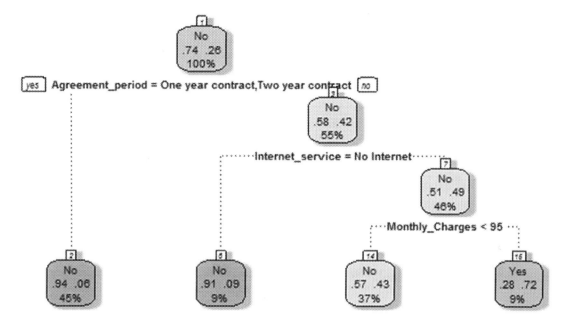

Figure 4-8. *Pruned decision tree*

```
# Predicting the model using test data

test_data$Churn_Class <- predict(prune_model,
newdata = test_data, type="class")

#Display the confusion matrix or classification table

table(test_data$Churn ,test_data$Churn_Class)
```

In Table 4-1 test_data$Churn is considered for the actual outcomes and test_data$Churn_Class is considered for the predicted outcomes.

Table 4-1. *Classification Table*

test_data$Churn_Class		
test_data$Churn	No	Yes
No	218	4
Yes	58	20

Diagonal values are correctly classified, hence Accuracy rate can be calculated as:

$$Accuracy\ rate = \frac{218+20}{218+4+58+20} = \frac{238}{300} = 0.79$$

Error rate or Misclassification error rate can be calculated as:

$$1 - Accuracy\ rate = 1 - 0.79 = 0.21$$

```
# Predicting the probability matrix  using test data

pred <- predict(prune_model, newdata = test_data, type="prob")

pred1 <- data.frame(pred)
```

Probability matrix is displayed in Table 4-2.

Table 4-2. *Customer Churn Proability Table*

Probability(No)	Probability%(No)	Probability(Yes)	Probability%(Yes)
0.942675	94.27%	0.057325	5.73%
0.569767	56.98%	0.430233	43.02%
0.942675	94.27%	0.057325	5.73%
0.90625	90.63%	0.09375	9.38%
0.569767	56.98%	0.430233	43.02%
0.942675	94.27%	0.057325	5.73%
0.28125	28.13%	0.71875	**71.88%**
0.28125	28.13%	0.71875	**71.88%**
0.942675	94.27%	0.057325	5.73%
0.942675	94.27%	0.057325	5.73%

Probability (No) represents the probability value of customers not churning and Probability% (No) represents the percentage of customers who will not churn, and Probability (Yes) represents the probability value of customer churn, and Probability% (Yes) represent the percentage of customers who will churn. All telecommunications industries are using these predicting probabilities to target the customers who are having a high likelihood to churn with the promotional offers, providing the best customer service and engagement, which results in reduction of customer churn. For example, the probability percentage of customers highlighted in bold font in Table 4-2 are in a risky zone and it signifies that these two customers are having 71.88% of probability to churn in the future and need to be targeted with improved customer service and offers in order to retain them.

Decision Tree Model Using SAS

In this section, we discuss about different SAS procedures like proc content, proc means, proc freq, and proc univariate. We also discuss about building a decision tree model for predicting the probability of customer churn with an explanation of SAS codes and output of each part in program 2 and program 2.1.

```
/*Create your own library in SAS like here it is libref and mention the
path */
libname libref "/home/aro1260/deep";

/*Importing churn_dataset in your assigned library */
PROC IMPORT DATAFILE= "/home/aroragaurav1260/data/churn_dataset.csv"
     DBMS=CSV Replace
        OUT=libref.churn;
        GETNAMES=YES;
RUN;

/* To check the contents of the data */
PROC CONTENTS DATA=libref.churn;
RUN;
```

The CONTENTS Procedure

Data Set Name	LIBREF.CHURN	**Observations**	1000
Member Type	DATA	**Variables**	14
Engine	V9	**Indexes**	0
Created	12/11/2017 14:20:53	**Observation Length**	112
Last Modified	12/11/2017 14:20:53	**Deleted Observations**	0
Protection		**Compressed**	NO
Data Set Type		**Sorted**	NO
Label			
Data Representation	SOLARIS_X86_64, LINUX_X86_64, ALPHA_TRU64, LINUX_IA64		
Encoding	utf-8 Unicode (UTF-8)		

Alphabetic List of Variables and Attributes

#	Variable	Type	Len	Format	Informat
11	Agreement_period	Char	17	$17.	$17.
14	Churn	Char	3	$3.	$3.
5	International_plan	Char	3	$3.	$3.
8	Internet_service	Char	11	$11.	$11.
2	Marital_Status	Char	7	$7.	$7.
12	Monthly_Charges	Num	8	BEST12.	BEST32.
7	Multiple_line	Char	8	$8.	$8.
4	Phone_service	Char	3	$3.	$3.
1	Sex	Char	6	$6.	$6.
10	Streaming_Videos	Char	11	$11.	$11.

(continued)

194

Alphabetic List of Variables and Attributes

#	Variable	Type	Len	Format	Informat
9	Technical_support	Char	11	$11.	$11.
3	Term	Num	8	BEST12.	BEST32.
13	Total_Charges	Num	8	BEST12.	BEST32.
6	Voice_mail_plan	Char	3	$3.	$3.

Proc content displays the content of the data like the number of observations, number of variables in the data, library name, and data type of each variable whether the variables are numeric or character with their Length, Format, and Informat.

```
/*Descriptive statistics of the data*/
proc means data = libref.churn;
var  Term Monthly_Charges;
run;
```

The MEANS Procedure

Variable	N	Mean	Std Dev	Minimum	Maximum
Term	1000	32.8050000	25.1356475	0	72.0000000
Monthly_Charges	1000	66.6393500	30.2986094	18.9500000	116.2500000

Proc means displays the descriptive statistic or summary of the data like Number of observations (N), Mean, Standard deviation, Minimum and Maximum values of the respective variable.

```
/* Applying Proc freq to see the frequency of the data */
Proc freq data = libref.churn;
tables Internet_service  Churn  Churn * Internet_service;
run;
```

The FREQ Procedure

Internet_service	Frequency	Percent	Cumulative Frequency	Cumulative Percent
Cable	171	17.10	171	17.10
DSL	280	28.00	451	45.10
Fiber optic	341	34.10	792	79.20
No Internet	208	20.80	1000	100.00

Churn	Frequency	Percent	Cumulative Frequency	Cumulative Percent
No	741	74.10	741	74.10
Yes	259	25.90	1000	100.00

Proc freq represents the number of frequencies of each level with cumulative frequency and cumulative percentage, like number of Cable, number of DSL, number of Fiber optic, number of No Internet, number of customers not churned (No) and number of customers churned (Yes).

Frequency Percent Row Pct Col Pct	Table of Churn by Internet_service				
	Churn	Internet_service			
		Cable	DSL	Fiber optic	No Internet	Total
	No	103	221	219	198	741
		10.30	22.10	21.90	19.80	74.10
		13.90	29.82	29.55	26.72	
		60.23	78.93	64.22	95.19	
	Yes	68	59	122	10	259
		6.80	5.90	12.20	1.00	25.90
		26.25	22.78	47.10	3.86	
		39.77	21.07	35.78	4.81	
	Total	171	280	341	208	1000
		17.10	28.00	34.10	20.80	100.00

Interaction between two variables like Churn and Internet_service display the table of Churn by Internet_service with frequency, percent, row percent, and column percent details.

```
/* Applying proc univariate to get more detailed summary of the data */

proc univariate data = libref.churn;
var Monthly_Charges;
histogram  Monthly_Charges/normal;
run;
```

The UNIVARIATE Procedure

Variable: Monthly_Charges

Moments

N	1000	Sum Weights	1000
Mean	66.63935	Sum Observations	66639.35
Std Deviation	30.2986094	Variance	918.00573
Skewness	-0.3099341	Kurtosis	-1.2342213
Uncorrected SS	5357890.69	Corrected SS	917087.724
Coeff Variation	45.466544	Std Error Mean	0.95812616

Basic Statistical Measures

Location		Variability	
Mean	66.63935	Std Deviation	30.29861
Median	74.72500	Variance	918.00573
Mode	20.05000	Range	97.30000
		Interquartile Range	50.77500

Note: The mode displayed is the smaller of 2 modes with a count of 12.

Tests for Location: Mu0=0

Test	Statistic		p Value	
Student's t	t	69.55175	Pr > ltl	<.0001
Sign	M	500	Pr >= IMI	<.0001
Signed Rank	S	250250	Pr >= ISI	<.0001

Quantiles (Definition 5)

Level	Quantile
100% Max	116.250
99%	115.025
95%	108.050
90%	104.475
75% Q3	90.900
50% Median	74.725
25% Q1	40.125
10%	20.350
5%	19.875
1%	19.350
0% Min	18.950

Extreme Observations

Lowest		Highest	
Value	Obs	Value	Obs
18.95	115	116.05	153
19.15	1000	116.10	354
19.15	802	116.10	527
19.15	721	116.10	922
19.20	729	116.25	877

Proc univariate is used to display the detailed summary or descriptive statistics of the data. It will display kurtosis, skewness, standard deviation, uncorrected ss, corrected ss, standard error mean, variance, range, interquartile range, etc.

It helps in assessing the normal distribution of the data, goodness-of-fit test for normal distribution and in detecting the outliers or extreme values present in the data. Figure 4-9 displays the normal distribution of the Monthly_Charges present in the data.

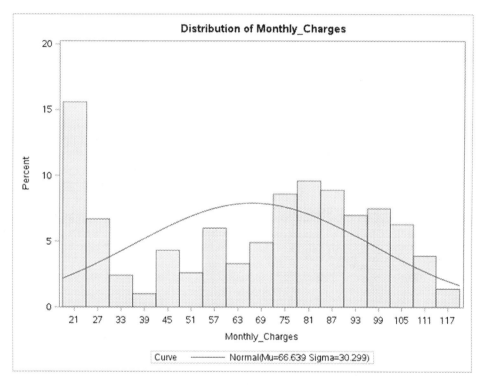

Figure 4-9. *Normal distribution for Monthly_Charges*

The UNIVARIATE Procedure

Fitted Normal Distribution for Monthly_Charges

Parameters for Normal Distribution

Parameter	Symbol	Estimate
Mean	Mu	66.63935
Std Dev	Sigma	30.29861

Goodness-of-Fit Tests for Normal Distribution

Test		Statistic	p Value	
Kolmogorov-Smirnov	D	0.1279262	Pr > D	<0.010
Cramer-von Mises	W-Sq	3.9534168	Pr > W-Sq	<0.005
Anderson-Darling	A-Sq	27.9980535	Pr > A-Sq	<0.005

Model Building and Interpretation of Full Data

The program2 code **ODS GRAPHICS ON** is used to create the plots that are the part of the output in the decision tree model. **HPSPLIT procedure** is used to create the classification tree. **CVMODELFIT** is used for model assessment and displaying the cross validation confusion matrix based on 10-fold cross validation. **SEED** option is used to wwspecify the random number for the cross validation process, in this case **seed = 123.**

    MAXDEPTH statement specifies the tree depth by default, and the growth process of the tree is 10 (**maxdepth value is 10),** but different limits can be specified according to the requirement and in this case study **depth = 3** is used to grow the decision tree till **depth 3.CLASS** statement is used to include all the variables that are categorical. **MODEL** statement signifies **Churn** as the dependent variable and to the right of the equal sign are all independent variables. **EVENTS** is used in binary categorical dependent variable to specify the event levels, **events = 'Yes'** is considered as event of interest as in this case study the modeling is done for the customers who have high likelihood to churn in the future.

    GROW statement is used to identify the splitting criterion in the decision tree for splitting the parent nodes into child nodes as the tree grows; by default the splitting criterion is **entropy** but other criterion like gini, variance reduction, Chi-Square, etc., can also be used; in this case study the splitting criterion of the decision tree is **grow entropy**.

    PRUNE statement is used to prune the trees as fully grown trees or large trees have overfitting issues that will impact the predictive accuracy of the tree in order to minimize the overfitting issues and prediction error; the large trees are pruned, and by default the pruning method is **cost complexity;** and in this case the study **prune costcomplexity** is used. **LEAVES** is used for specifying the number of leaves used in the tree – in this case study **leaves = 4** is used.

/* Growing Decision tree on Full data*/

Program2:

```
ods graphics on;
proc hpsplit data=libref.churn cvmodelfit seed=123 maxdepth=3;
class Churn Sex    Marital_Status Phone_service International_plan Voice_
mail_plan Multiple_line Internet_service Technical_support Streaming_Videos
Agreement_period;
model Churn (event='Yes') = Sex     Marital_Status
Term    Phone_service International_plan    Voice_mail_plan Multiple_line
Internet_service Technical_support Streaming_Videos
Agreement_period Monthly_Charges Total_Charges;
grow entropy;
prune costcomplexity(leaves=4);
run;
```

The **decision tree output of program2** is split into several sections and each is discussed in the following section.

Performance Information table is displayed in Part 1 of program 2, which describes that the HPSPLIT procedure executes in single machine mode and the number of threads is equal to 2.

Part 1

The HPSPLIT Procedure

Performance Information

Execution Mode	Single Machine
Number of Threads	2

Data Access Information is displayed in Part 2 of program2, which describes that the input data set is accessed using the V9 base engine on the client machine where the SAS session executes.

Part 2

Data Access Information			
Data	**Engine**	**Role**	**Path**
LIBREF.CHURN	V9	Input	On Client

Model Information is displayed in Part 3 of program 2, which delivers the model information and the methods used to grow and prune the decision tree-like Splitting criterion is Entropy, Pruning method is Cost complexity, Subtree Evaluation Criterion is Number of leaves, Number of branches is 2, Maximum tree depth requested is 3, Maximum tree depth achieved is 3, Tree depth is 3, Number of leaves before pruning is 8, Number of leaves after pruning is 4, and Model event level is Yes.

Part 3

Model Information	
Split Criterion Used	Entropy
Pruning Method	Cost Complexity
Subtree Evaluation Criterion	Number of Leaves
Number of Branches	2
Maximum Tree Depth Requested	3
Maximum Tree Depth Achieved	3
Tree Depth	3
Number of Leaves Before Pruning	8
Number of Leaves After Pruning	4
Model Event Level	Yes

Part 4 of program 2 displays the information about the Number of observations read is 1,000 and Number of observations used is 1,000. In this case study there are no missing values so Number of observations read and used are the same.

Part 4

Number of Observations Read	1000
Number of Observations Used	1000

Part 5 and Part 6 of program 2 displays the 10-Fold Cross Validation Assessment of model and 10-Fold Cross Validation Confusion Matrix.

Part 5

The HPSPLIT Procedure

10-Fold Cross Validation Assessment of Model

N Leaves	Average Square Error				Misclassification Rate			
	Min	Avg	Standard Error	Max	Min	Avg	Standard Error	Max
4	0.1334	0.1563	0.0186	0.2040	0.2056	0.2468	0.0301	0.3053

Part 6

10-Fold Cross Validation Confusion Matrix

Actual	Predicted		Error Rate
	No	Yes	
No	684	57	0.0769
Yes	189	70	0.7297

Part 7 of program 2 is displayed in Figure 4-10. It displays the overview of the complete decision tree. Each leaf node has a color bar that specifies the most frequent level of churn and represents the classification level allocated to all observations in that particular node. The bar height represents the fraction of observations (churn) in that particular node having the most frequent level. In this case study No (customers not churn) is the most frequent level as compared to Yes (customers churn)

Part 7

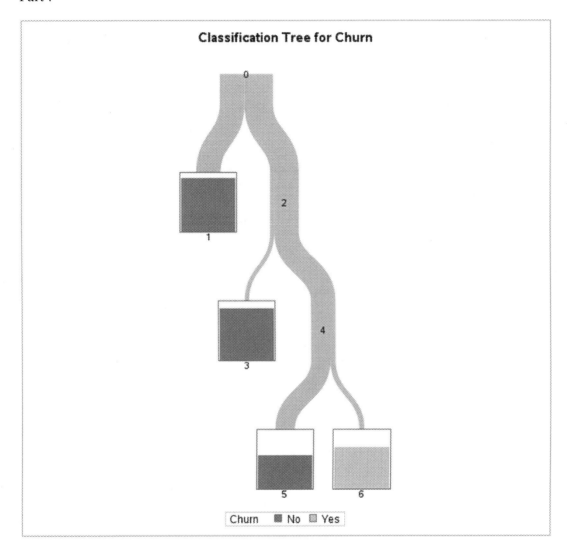

Figure 4-10. *Program 2 Overview of decision tree*

Part 8 of program 2 is displayed in Figure 4-11. It displays the details of the complete decision tree. The first split is based on **Agreement_period**, and it means **Agreement_period** is considered a highly significant variable in the model.

There are **1,000** customers from churn_dataset where in root node **74%** of customers do not churn and **26%** customers churn. Customers for whom **Agreement_period= Monthly contract**, the customer churn is **42.49%.** The customers for whom **Agreement_period= Monthly contract** are further subdivided based on the **Internet_service** variable. There are **48.25%** of the customer churn for whom **Internet_service=Cable,**

DSL, Fiber optic. The customers for whom **Internet_service=Cable, DSL, Fiber optic** are further subdivided on the basis of **Monthly_Charges** variable. There are **72.16%** of the customers churn for whom **Internet_service=Cable, DSL, Fiber optic** and **Monthly_Charges is >= 94.844**. Contrast to these customers for whom **Internet_service = No**, there is **9.88%** of customer churn. In summary, based on the decision tree splits, customers who have a high likelihood of churn are those for whom **Internet_service=Cable, DSL, Fiber optic and Monthly_Charges is >= 94.844**.

Part 8

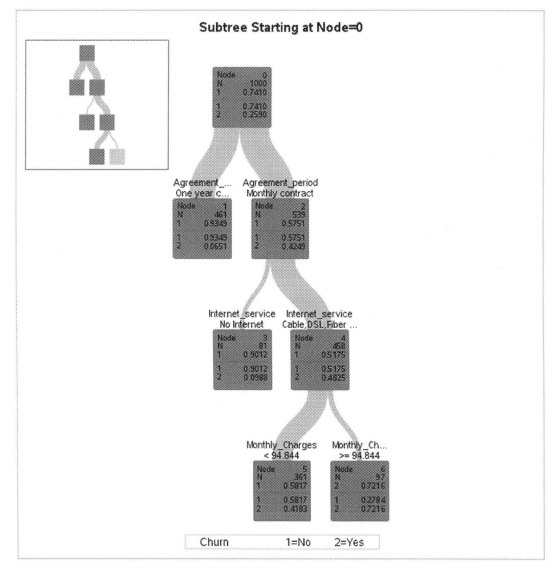

Figure 4-11. *Program 2 Detailed plot of decision tree*

Part 9 of program 2 displays the two Confusion Matrices, Model Based and Cross Validation Confusion Matrix. and it is used for estimating the accuracy of the decision tree. A model-based confusion matrix is formed by applying the fitted model to the training data and cross validation matrix is based on a 10-fold cross validation and is formed by using CVMODELFIT option in the code.

Part 9

The HPSPLIT Procedure

Confusion Matrices

	Actual	Predicted		Error Rate
		No	Yes	
Model Based	No	714	27	0.0364
	Yes	189	70	0.7297
Cross Validation	No	684	57	0.0769
	Yes	189	70	0.7297

Part 10 of program 2 displays the Fit Statistics for Selected tree, and there are two fit statistics – one denotes the model based and the other one denotes the cross validation based on 10-fold cross validation. The model-based misclassification rate is **21.60%** and cross validation misclassification rate is **24.68%**, which is higher than the model-based misclassification rate. The Model-based Area under curve or AUC is **79.77%**, which indicates that the overall predictive accuracy of the model is good.

Part 10

Fit Statistics for Selected Tree

	N Leaves	ASE	Mis- class	Sensitivity	Specificity	Entropy	Gini	RSS	AUC
Model Based	4	0.1426	0.2160	0.2703	0.9636	0.6345	0.2852	285.2	0.7977
Cross Validation	4	0.1563	0.2468	0.2703	0.9231				

Part 11 of program 2 is displayed in Figure 4-12. It displays the ROC curve for the Churn binary dependent variable and summarizes the performance of the decision tree model. The higher the AUC or ROC curve value, the better is the predictive accuracy of the model.

Part 11

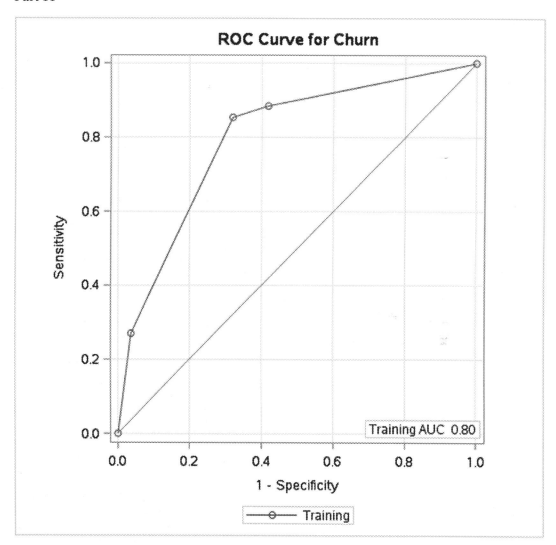

Figure 4-12. *Program 2 ROC plot*

In this case study Part 12 of program 2 displays the Variable Importance table, which specifies the most important independent variable used for predicting the dependent variable customer churn. There are various types of metrics for measuring the variable's importance in the model like count, RSS, and relative importance.

Part 12

Variable Importance

Variable	Training		Count
	Relative	Importance	
Agreement_period	1.0000	8.0205	1
Internet_service	0.5614	4.5026	1
Monthly_Charges	0.4677	3.7514	1

Model Building and Interpretation on Training and Testing Data

In program 2.1 the code HPSPLIT procedure is used to create the classification tree. The Plots = zoomedtree option in the statement is used to call diagrams that begin with other nodes like in this case nodes = 0; it means call diagrams that begin with 0, and if nodes = 2, then call diagrams that begin with 2, and so on. The DEPTH statement specifies the tree depth by default and the growth process of the tree is 10 (maxdepth value is 10), but different limits can be specified according to the requirement, and in this case depth = 3 is used to grow the decision tree until depth 3. CLASS statement is used to include all the variables that are categorical. The MODEL statement signifies Churn as the dependent variable and to the right of the equal sign are all independent variables.

EVENTS is used in the binary categorical dependent variable to specify the event levels, events = 'Yes' is considered as event of interest as in this case the modeling is done for the customers who have a high likelihood to churn in the future. GROW statement is used to identify the splitting criterion in the decision tree for splitting the parent nodes into child nodes as the tree grows; by default the splitting criterion is entropy but other criterion like gini, variance reduction, Chi-Square, etc., can also be used, in this case splitting criterion of the decision tree is grow entropy.

PRUNE statement is used to prune the trees as fully grown trees or large trees have overfitting issues that will impact the predictive accuracy of the tree in order to minimize the overfitting issues and prediction error. The large trees are pruned, and by default the pruning method is cost complexity; in this case prune cost complexity is used. LEAVES is used for specifying the number of leaves used in the tree and in this case, the study leaves = 6 means that the tree with 6 leaves is selected.

PARTITION STATETEMENT is used to split the observations in the churn_dataset logically into training and validation datasets; in this case, the study partition fraction(validate=0.3 seed=123) is used for splitting the dataset as validation dataset observations are selected on a random basis with probability of 0.3, and the remaining probability of 0.7 will be selected for the training dataset. SEED option is used to specify the random number for cross validation process, in this case seed = 123. FILE = OPTION in the CODE statement is used to save the SAS DATA step score code in the file named as scorefile.sas and the FILE = OPTION in the RULES statement is used to save the node rules in the file named as noderules.txt.

```
/* Growing Decision tree and splitting data into trainig and validation*/
```

Program2.1:

```
ods graphics on;
proc hpsplit data=libref.churn
plots=zoomedtree(nodes=('0') depth=3);
class Churn Sex    Marital_Status Phone_service International_plan Voice_
mail_plan Multiple_line Internet_service Technical_support Streaming_Videos
Agreement_period;
model Churn (event='Yes') = Sex    Marital_Status
Term    Phone_service International_plan    Voice_mail_plan Multiple_line
Internet_service Technical_support Streaming_Videos    Agreement_period
Monthly_Charges Total_Charges;
grow entropy;
prune costcomplexity(leaves=6);
partition fraction(validate=0.3 seed=123);
code file= "/home/aroragaurav1260/data/scorefile.sas";
rules file="/home/aroragaurav1260/data/noderules.txt";
run;
```

Note that

The **decision tree output of program 2.1** is split into several sections and each of is discussed in the following section.

Parts 1, 2, and 3 output of program 2.1 is explained previously for program 2.

Part 1

The HPSPLIT Procedure

Performance Information

Execution Mode	Single Machine
Number of Threads	2

Part 2

Data Access Information

Data	Engine	Role	Path
LIBREF.CHURN	V9	Input	On Client

Part 3

Model Information

Split Criterion Used	Entropy
Pruning Method	Cost Complexity
Subtree Evaluation Criterion	Number of Leaves
Number of Branches	2
Maximum Tree Depth Requested	10
Maximum Tree Depth Achieved	10
Tree Depth	5
Number of Leaves Before Pruning	96
Number of Leaves After Pruning	6
Model Event Level	Yes

Part 4 of program 2.1 displays the number of observations read and used, which is equal to 1,000. The training dataset has a random 709 observations, used for model building; and the validation dataset is has a random 291 observations, used for model validation.

Part 4

Number of Observations Read	1000
Number of Observations Used	1000
Number of Training Observations Used	709
Number of Validation Observations Used	291

Part 5 of program 2.1 is displayed in Figure 4-13 and its output is explained previously for Part 7 of program 2.

Part 5

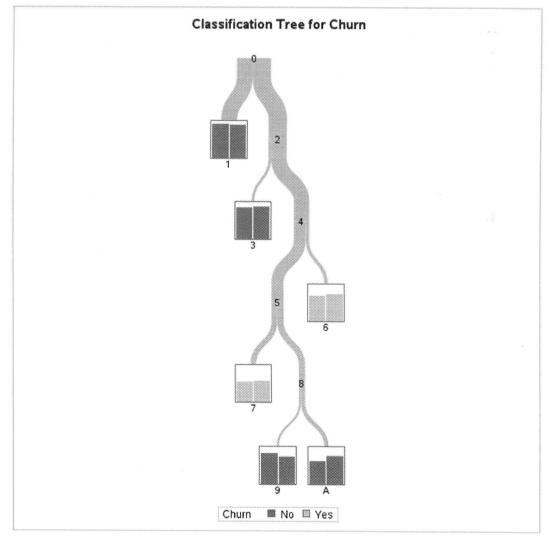

Figure 4-13. *Program 2.1 Overview of decision tree*

Part 6 of Program 2.1 is displayed in Figure 4-14. It displays the details of the complete decision tree. The first split is based on **Agreement_period**, and it means the **Agreement_period** is considered a highly significant variable in the model.

There are **709** customers from training churn_dataset where in root node **74%** of customers do not churn and **26%** customers churn. Customers for whom **Agreement_period= Monthly contract**, the customer churn is **43.01%.** The customers for whom **Agreement_period= Monthly contract** are further subdivided based on the **Internet_service** variable. There is a **48.76%** of the customer churn for whom **Internet_service=Cable, DSL, Fiber optic**. The customers for whom **Internet_service=Cable, DSL, Fiber optic** are further subdivided on the basis of **Monthly_Charges** variable. There is a **70.83%** customers churn for whom **Internet_service=Cable, DSL, Fiber optic** and **Monthly_Charges is >= 94.844**. Contrast this to customers for whom **Internet_service = No**, there are **10.53%** of the customer churn. In summary, based on the decision tree splits, customers who have a high likelihood of churn are those for whom **Internet_service=Cable, DSL, Fiber optic and Monthly_Charges is >= 94.844**.

Part 6

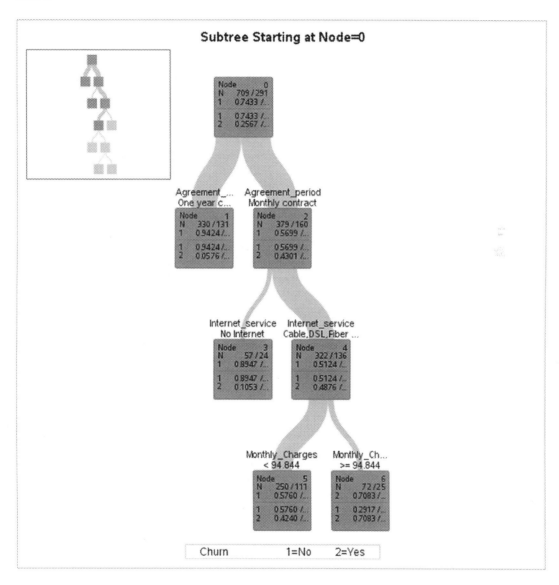

Figure 4-14. *Program 2.1 Detailed plot of decision tree*

Part 7 of program 2.1 and its output is explained previously for program 2.

Part 7

The HPSPLIT Procedure

Confusion Matrices

	Actual	Predicted		Error Rate
		No	Yes	
Training	No	457	70	0.1328
	Yes	63	119	0.3462
Validation	No	186	28	0.1308
	Yes	26	51	0.3377

Part 8 of program 2.1 and its output are explained previously for program 2. In this case, an additional thing is Fit Statistics for the selected tree and is based on Training and Validation. In training, the misclassification is **18.76%** and AUC is **83%** while in validation, misclassification is **18.56%** and AUC is **81%.**

Part 8

Fit Statistics for Selected Tree

	N Leaves	ASE	Mis- class	Sensitivity	Specificity	Entropy	Gini	RSS	AUC
Training	6	0.1305	0.1876	0.6538	0.8672	0.5924	0.2610	185.0	0.8344
Validation	6	0.1393	0.1856	0.6623	0.8692	0.6224	0.2718	81.0487	0.8110

Part 9 of program 2.1 is displayed in Figure 4-15 and its output is explained previously for program 2.

Part 9

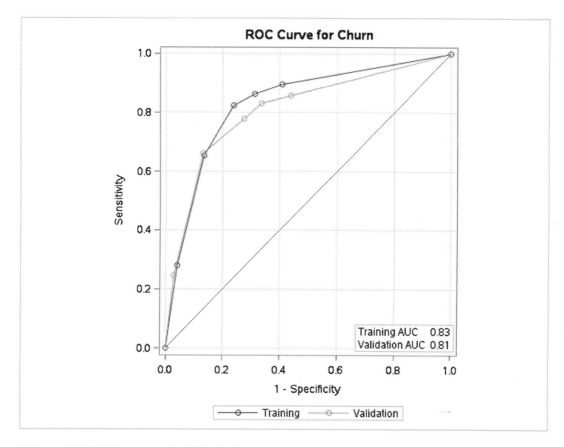

Figure 4-15. *Program 2.1 ROC plot*

Part 10 of program 2.1 is displayed below and its output is explained previously for program 2.

Part 10

Variable Importance

Variable	Training		Validation		Relative Ratio	Count
	Relative	Importance	Relative	Importance		
Agreement_period	1.0000	6.9968	1.0000	3.9119	1.0000	1
Term	0.4712	3.2968	0.6789	2.6558	1.4408	1
Internet_service	0.5378	3.7625	0.6322	2.4730	1.1756	1
Monthly_Charges	0.4297	3.0064	0.5716	2.2359	1.3302	1
Streaming_Videos	0.2263	1.5837	0.0000	0	0.0000	1

/*scoring the data to predict the probability of customer churn*/

```
data libref.finalscore;
set libref.churn;
%include "/home/aro1260/data/scorefile.sas";
run;
```

A probability table of customer churn is displayed in Table 4-3. V_ChurnNo represents the probability value of customer not churn and V_ChurnNo% represents the percentage of customers who will not churn from the validation dataset and V_ChurnYes represents the probability value of customer churn and V_ChurnYes% represents the percentage of customers who will churn from the validation dataset. With the help of the decision tree model, the probability of customer churn can be predicted and helps in reducing customer churn. For example, the probability percentage of customers highlighted in bold font in Table 4-3 are in a risky zone and it signifies that these three customers have a high probability of churning in the future and need to be targeted with improved customer service and promotional offers in order to retain them.

Table 4-3. *Customer Churn Proability Table in SAS*

V_ChurnNo	V_ChurnNo %	V_ChurnYes	V_ChurnYes%
0.24	24.0%	0.76	**76.0%**
0.91603053	91.6%	0.083969466	8.4%
0.76470588	76.5%	0.23529412	23.5%
0.91603053	91.6%	0.083969466	8.4%
0.24	24.0%	0.76	**76.0%**
0.91666667	91.7%	0.083333333	8.3%
0.40740741	40.7%	0.59259259	59.3%
0.76470588	76.5%	0.23529412	23.5%
0.24	24.0%	0.76	**76.0%**
0.91603053	91.6%	0.083969466	8.4%

Summary

In this chapter we learned about different data analytics applications in the telecommunications industry. We also learned about the decision tree model and its various characteristics and features. The case study on predicting the probability of customer churn in the telecommunications industry demonstrated a practical application of this model in a real-life scenario. We also mastered how to develop a model, execute, visualize, and interpret results using both R and SAS Studio.

References

1. Huurdeman, A. A., *The Worldwide History of Telecommunications*. Wiley: 2003.

2. Bertin, E.; Crespi, N.; Magedanz, T., *Evolution of Telecommunication Services: The Convergence of Telecom and Internet: Technologies and Ecosystems*. Springer Berlin Heidelberg: 2013.

3. Parietti, M. The World's Top 10 Telecommunications Companies: 2017.

4. Anttalainen, T., *Introduction to Telecommunications Network Engineering*. Artech House: 2003.

5. Perrucci, M. R. a. A. *Churn Analysis Case Study*; Telecom Italia Lab: 2002; p. 12.

6. Deloitte *Opportunities in Telecom Sector: Arising from Big Data*; Aegis: 2015; p. 32.

7. Express, F. How Verizon Cut Customer Churn By 0.5 Per Cent *Tech Verticals* [Online], 2003.

8. Jim Wheless, M. G. New Generation of Network Analytics: Network Optimization for Communications Service Providers: 2013.

9. Herrera, C. A. A.; Özdemir, D.; Cabrera-Ríos, M., Capacity Planning In a Telecommunications Network: A Case Study. *International Journal of Industrial Engineering: Theory, Applications and Practice* **2009,** *16* (2), 9.

10. Sousa, J. V. C. d. Telecommunication Fraud Detection Using Data Mining techniques. University of Porto, 2014.

11. Luis Cortesão, F. M., António Rosa, Pedro Carvalho, Fraud Management Systems in Telecommunications: a practical approach. In *International conference on telecommunications*, IEEE: Capetown, South Africa, 2005.

12. Kurtz, N., Securing A Mobile Telecommunications Network From Internal Fraud. SANS Institute: 2002.

13. Kuşaksızoğlu, B. Fraud Detection in mobile communication networks using data mining. The University of Bahcesehir, Turkey, 2006.

14. Estévez, P. A.; Held, C. M.; Perez, C. A., Subscription fraud prevention in telecommunications using fuzzy rules and neural networks. *Expert Systems with Applications* **2006,** *31* (2), 337–344.

15. Robert Phillips, F. R., Pricing and Revenue Optimization. *The RMA Journal* **2006**, 4.

16. Bondos, I., Price and service bundling- An example of the telecommunications market in Poland. In *Human capital without borders: Knowledge and learning for Quality of life*, Portoroz, Slovania, 2014; p 10.

17. Zhou, Z. H., *Ensemble Methods: Foundations and Algorithms*. Taylor & Francis: 2012.

18. Kingsford, C.; Salzberg, S. L., What are decision trees? *Nature biotechnology* **2008,** *26* (9), 1011–1013.

19. Sachin Gavankar, S. S., Decision Tree: Review of Techniques for Missing Values at Training, Testing and Compatibility. In *Third International Conference on Artificial Intelligence, Modelling and Simulation*, IEEE Computer Society: Sabah, Malaysia, 2015; pp. 122–126.

20. Saar-Tsechansky, M.; Provost, F., Handling Missing Values when Applying Classification Models. *J. Mach. Learn. Res.* **2007,** *8*, 1623–1657.

21. **Bramer**, M., Avoiding Overfitting of Decision Trees. In *Principles of Data Mining*, Springer London: London, 2007; pp. 119–134.

22. Maimon, O.; Rokach, L., *Data Mining and Knowledge Discovery Handbook*. Springer US: 2006.

23. Raileanu, L. E.; Stoffel, K., Theoretical Comparison between the Gini Index and Information Gain Criteria. *Annals of Mathematics and Artificial Intelligence* **2004,** *41* (1), 77–93.

24. Rutkowski, L.; Jaworski, M.; Pietruczuk, L.; Duda, P., The CART decision tree for mining data streams. *Information Sciences* **2014,** *266* (Supplement C), 1-15.

25. Badr Hssina, A. M., Hanane Ezzikouri and Mohammed Erritali, A comparative study of decision tree ID3 and C4.5. *International Journal of Advanced Computer Science and Applications(IJACSA), Special Issue on Advances in Vehicular Ad Hoc Networking and Applications* **2014**, pp. 13–19.

26.   Geurts, P.; Wehenkel, L., *Investigation and Reduction of Discretization Variance in Decision tree Induction*. 2000.

27.   Kitts, B., Regression Trees. `http://www.appliedaisystems.com/papers/RegressionTrees.pdf`, 1999; pp. 1–15.

CHAPTER 5

Healthcare Case Study

The healthcare industry is growing at a very fast pace with global healthcare spending projected to touch USD 8.7 trillion in 2020.[1] The healthcare sector comprises the industries that specialize in products and services that provide health and medical care to patients. The sector can be broadly classified under four main subsectors displayed in Figure 5-1.

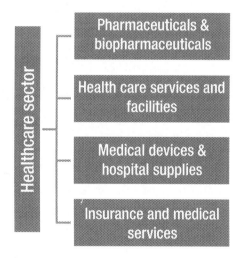

Figure 5-1. *Subsectors in healthcare sector*

Pharmaceutical and biopharmaceutical subsectors consist of industries involved in research, development, and manufacturing of therapeutics. Biopharmaceutical segment continues to grow at a tremendous pace of approximately 15% with annual sales touching USD $200 billion in 2016.[2] Healthcare services consist of a broad array of services provided by healthcare professional like doctors, nurses, medical technicians, etc., for the patients. The services can fall under hospital and dental services and intermediate and long-term care. Healthcare facilities are categorized under hospitals, ambulatory surgical centers, doctors' offices, urgent care centers, and nursing homes based on capabilities, specialization, and available services. Medical Devices is also a

© Deepti Gupta 2018
D. Gupta, *Applied Analytics through Case Studies Using SAS and R*,
https://doi.org/10.1007/978-1-4842-3525-6_5

big segment. In 2015, the global medical device market was approximately USD $350 billion and has seen a steady growth.[3] Last but definitely not least, medical insurance has evolved into a huge industry. In 2017, the total global revenue from the private health insurance sector was USD $1.59 trillion.[4] Other medical services include industries like healthcare staff recruitment agencies, medical apparel manufacturing, medical waste disposal services, medical couriers, etc. Like many other sectors, technology, data analytics, and digitization had a transformatory impact on every aspect of the industry. Medical research is employing robotics, advanced analytical methods and imagery, and data analytics to accelerate and optimize drug discovery and development.[5, 6] Advanced diagnostics coupled with data analytics enable doctors to accurately identify ailments at an early stage as well as identify the systemic trends and variables contributing to causes of the ailment.[7] Modern medical devices help deliver precise and targeted therapeutics,[8] and the biotechnology and biopharmaceutical revolution helped realize cures for various chronic illnesses.[9] Advances in electronic medical records and healthcare insurance have streamlined logistics and reduced fraud.[10]

The healthcare industry has changed dramatically during the last 25 years, and there are several changes that have occurred in the healthcare industry and will continue to evolve over the coming years. The major transitions observed in the healthcare industry in terms of financial management perspective are listed in the following section.[11]

1. **The shift of services from inpatient to outpatient**

    Medical treatment and surgeries can be categorized into inpatient and outpatient procedures. Inpatient care results in overnight hospitalization for the treatment and surgery while in outpatient the medical care and treatment are provided to the patients without spending a single night in the hospital. The highest percentage of the revenue generated by the hospitals is due to the outpatient services, and the drivers responsible for the shift from inpatient to outpatient services are because of the advanced and improved technology, payment incentives, and demand from the patients. Advance and improved technology has made many inpatient procedures transition to outpatient. The four prominent ones include hysterectomy, angioplasty, gall bladder removal, and lumbar surgery.[12] Patients always opt for outpatient care if they feel that there are not any side effects of the procedures as they are less expensive than inpatient care and time saving.

2.   Transition to electronic healthcare records

Traditional paper-based records were not easily accessible and doctors and other healthcare professionals had to rely upon documentation diligence of the patients for their medical history and availability of the records with them at all times. Digitization of medical records has significantly improved transparency, accessibility, reduced loss of information in transition, and reduced fraud. Additionally, electronic records have significantly higher survivability during disasters and natural calamities.[13] They also enable the possibility of better disaster response during these natural calamities.[14] Like all technologies there are some drawbacks of electronic medical records, which include time invested by medical practitioners to fill the records, cost of managing the records, and vulnerability to hacking. Overall the advantages significantly outweighs the drawbacks.[15]

3.   Increasing importance on value-based care with less cost

In recent years healthcare has transitioned to a full-scale financial industry. A major change that came about with this transition is the financial accountability of the institution. Traditionally the healthcare system gave limited emphasis on costs associated with the service. Uncontrollable increases in the cost of healthcare made the government and medical care providers introduce regulations and measures to cut down superfluous spending. Today's healthcare services are managed as a value-based service with a stringent cost monitoring. Technologies like robotics, laparoscopic methods, and pen devices(self-injecting syringes)[16] are becoming popular to mitigate cost and improve quality.

In this chapter the key applications of analytics that are reshaping the healthcare industry are discussed. Introduction to the Random Forest model and associated theoretical properties is provided. Application of the Random Forest in the healthcare industry has been demonstrated as a case study to predict the probability of malignant and benign breast cancer in R and SAS Studio.

Application of Analytics in the Healthcare Industry

The healthcare industry produces a huge amount of data on a daily basis. Healthcare data is gathered from different data sources like Health surveys, Electronic health records (EHR), Pharmacy, Diagnostic instruments, Laboratory information management systems, Insurance claims, and billing. Figure 5-2 displays the different data sources in the healthcare industry.

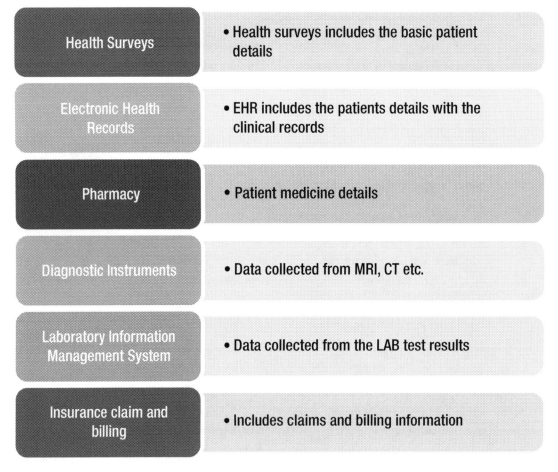

Figure 5-2. *Data sources in healthcare industry*

The healthcare industry is facing a big challenge in terms of managing the data produced on a daily basis in different formats like text, images, etc., by using traditional methods due to the shear high volume and variety in the data. Big data technologies are being implemented for storage, accessibility, security, and overall management of this vast amount of data. This vast amount of data offers an opportunity to employ predictive analytics and machine learning algorithms like logistic regression, decision tree, artificial neural network, and Random Forest, etc., to effectively withdraw insights and translate them into usable information. Analytics is playing an important role in redefining the healthcare industry in a holistic sense with applications ranging from predicting the outbreak of disease and preventative management, predicting the readmission rate of the patients, healthcare fraud detection, improving patient satisfaction, and identifying trends based on demographics. An overview of some of these key applications is provided in this section.

Predicting the Outbreak of Disease and Preventative Management

Analytics plays a significant role in predicting the likelihood of the outbreak of pandemics. Historical data of administrative and clinical information of the patients taken from electronic medical records, social media, and search engines are analyzed.[17] Predictive modeling and machine learning algorithms like the decision tree, neural network, etc., helps in analyzing the patterns of disease, identifying the origin, predicting the likelihood of outbreak, and demographics of patients.[18] Such tools can significantly improve preparedness and management of such pandemics. For example, finding the origin of the pandemic can help isolate the correct strain for vaccine development and by accurately quantifying the extent of outbreak, a good estimate of the number of vaccine doses that will need to be produced can be identified. These techniques can also help in identifying which regions or geographies need to be focused upon when a pandemic outbreak occurs.

For example, a Dengue outbreak prediction will send out early signals of the outbreak, which will be very helpful in identifying the location, time, duration, availability of the necessary facilities, and the allocation of the resources so the proper management can be done effectively and thousands of lives can be saved by providing the right treatment at the right time to the affected population.

Predicting the Readmission Rate of the Patients

A high readmission rate of patients not only increases the cost for the healthcare in the United States, but it also impacts the quality of the care for patients.[19] The Affordable Care Act (ACA) was signed into law in 2010, where high focus is given to the avoidable hospital readmission rates. Penalties and less payment is provided to the hospitals having a higher number of readmission rates and readmission of the patients in less than 30 days. On the one hand incentives are provided to the hospitals resulting in reduction in the number of avoidable readmission rates and quality care of the patients, hence reducing the operation costs of the hospital. In some cases the readmissions are unplanned, for example, in cancer treatment where the likelihood of unplanned 30-days cancer patients' readmissions are possible because of the progress of disease, obstacles in procedures, and new diagnoses.[19] The major drivers in increasing the readmission rate are inability to follow up with primary care, inability to refill and take medicines on time, lack of family support, inability to understand the discharge plan of care properly, and lack of awareness about health and effective lifestyle tips. Figure 5-3 displays the drivers of readmission rates in hospitals.

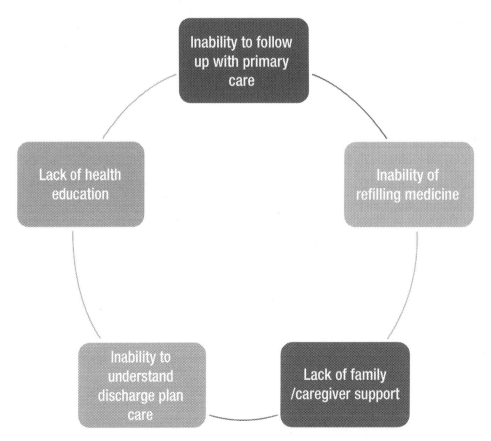

Figure 5-3. Factors responsible for high readmission rates in hospitals

Due to the availability of the data in Electronic Health Records (EHR), the patient information can be easily shared. With the help of predictive modeling and machine learning algorithms like the artificial neural network, decision tree, Random Forest, etc., can be applied to the patient historical data in order to predict the patients who have a high likelihood of getting readmitted to the hospital in less than 30 days. Advance predictions and early warning signals identify the actions or factors that may trigger an adverse impact. Predictive analytics helps in providing quality care to the high-risk patients on priority by providing outpatient follow-up, a physician appointment, and medication adherence intervention that will mitigate the risk of hospitals' high readmission rates.[19]

Healthcare Fraud Detection

One major challenge that the healthcare industry is facing is identification of fraudulent activities.[20] Cost efficient and effective healthcare is a societal need. Healthcare frauds are increasing the cost of healthcare services, which is impacting the end users as well as the government. The scale of the problem is continuously increasing, making it one of the top priorities for the governing bodies to identify sources of fraud and develop strategies to mitigate the occurrence.

There are various ways by which healthcare providers and patients can commit the healthcare fraud; a few of the common healthcare frauds committed by providers and policyholders are mentioned in the following section, for example.[21]

- Healthcare providers are billing for services that are not provided to the patients.

- Healthcare providers produce duplicate bills for the same service delivered to the patients.

- Healthcare providers generate the bill for non-covered services as covered services.

- Healthcare providers give more unnecessary tests and treatment to the patients that are not medically required, and at times the services provided are medically unrelated.

- Healthcare providers are charging more fees for the same services delivered to the patients as compared to their competitors.

- Healthcare providers misrepresenting the dates of the service provided to the patients.

- Healthcare providers misrepresenting the location of the service provided to the patients.

- Healthcare providers involving inincorrect reporting of diagnoses.

- Policyholders indulging in prescription drug fraud and false travel fraud.

- Policyholders giving their healthcare card access or letting others use their healthcare card in return for financial gains.

Traditional methods like account auditing are time consuming and not effective enough in detecting the fraudulent activities. With the help of data analytics, methods like the neural network, decision tree, and clustering techniques are useful in mining the huge chunks of healthcare industry data. The effects of extreme fraudulent claims are excessive amount of bills, higher costs of per patient, higher numbers of per-doctor patients, higher number of per-patient test, and many more. By using administrative and clinical information of the patients taken from electronic medical records and applying predictive modeling and machine learning techniques in that historical data helps in identifying the potential fraudulent claims and providers in advance, hence framing the strategies to prevent all fraudulent activities and save the healthcare industry from heavy financial losses.

Here are some real-time cases where fraud activities in the healthcare industry were responsible for huge financial losses: for example, 243 people were arrested, including doctors and nurses for $712 million Medicare fraud,[22] and the United States charged 412 people, including doctors, for $1.3 billion healthcare fraud, etc.[23].

Improve Patient Outcomes & Lower Costs

The next challenge faced by the healthcare analyst is to mine huge chunks of structure and unstructured electronic health records (EHR) data collected from multiple sources. By combining it with predictive analytics and machine learning techniques. it can be transformed into useful insights and provide the efficient healthcare services with lower costs. The Affordable Care Act came into law on March 23, 2010.[24] A goal of the Affordable Care Act (ACA) is to provide value-based care to the patients by providing efficient and improved healthcare quality services with lower costs.[25] Figure 5-4 displays the process for improved patient outcomes.

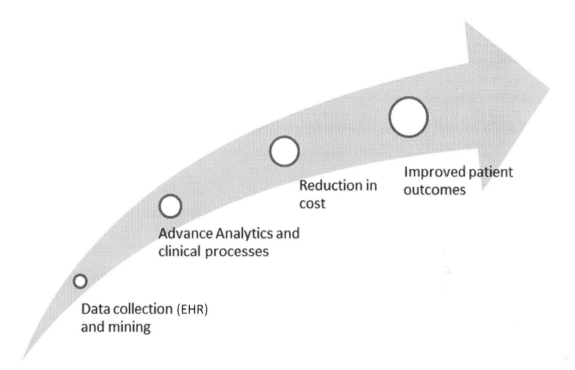

Figure 5-4. *Process for improved patients outcomes*

With the help of machine learning algorithms, the real-time monitoring of the outpatient is possible. The data from these monitors can be analyzed in real time and the alerts are sent to the healthcare providers continuously so they can be aware of any changes in the conditions of the patients and make the prompt decision to save the life of the patients by providing efficient treatment at the right time with lower costs. Predictive Analytics also helps in lowering the operation cost in the healthcare industry by properly utilizing the hospital's equipments and devices like there should be only one Xray machine, CT scan, etc., for each department, cutting off the unnecessary staff and nurse working hours for those days of the week when there is low flow of the patients.

Another important role of predictive analytics is to provide smooth patient flow and effective allocation of the resources. With the use of predictive analytical techniques, the patient flow in the healthcare industries can be predicting like which day of the week there is high flow in the number of patients. It will help healthcare management to take the necessary steps and effective management of resources by allocating more staff, nurses, number of beds, and doctors on those particular days, which will help in reducing the waiting time of the patients by providing them efficient services on time with lower cost.

Case Study: Predicting Probability of Malignant and Benign Breast Cancer with Random Forest Model

Random Forest is an ensemble learning method that consists of multiple decision trees, which are used to define the final output.[26] Random Forest is one of the most popular and powerful machine learning algorithms and one of the best among classification algorithms. The Random Forest algorithm was developed by Leo Breiman and Adele Cutler. Random Forest is used for classification and regression problems. For classification problems, the ensemble of the simple decision trees vote for the most popular class, and in regression problems, the estimate of the dependent variable is achieved by taking the average of their response. Ensemble learning is the method where weak learners combine together to form strong learners; therefore ensemble decision tree models will have the better predicting capability and higher accuracy. In Random Forests, each tree is grown to the maximal size; it means there is no pruning done in the trees – all trees are left unpruned.

A decision tree is built by using the complete data with all the features or variables whereas in Random Forest, each decision tree is grown based on the random selection of sample (with replacement) and random selection of a subset of features. Single decision trees have high variance issues but Random Forest helps in solving the high variance issues by averaging the variance, hence increasing the predictive accuracy of the model and eventually avoiding overfitting.

Working of Random Forest Algorithm

Random Forest grows many simple decision trees and each decision tree is grown as following manner.[26]

Step 1: Randomly select n number of sample cases with replacement from the N original data, ideally n< N. This sample n number of cases will be the training set used for growing the decision tree (where N is original data set and n is random sample or subset of the original data set).

Step 2: Randomly select p features from P features, where ideally p<<P and while growing Random Forest the value of p is held constant. Where P is total number of features or variables, p is a random subset of the features or variables.

(Leo Brieman suggests three possible values for p: $\sqrt{P}, 2\sqrt{P}, \frac{1}{2}\sqrt{P}$ for classification problems and $\frac{P}{3}$ for regression problems.)

Step 3: Next step is to calculate the best split among the p features based on different splitting criteria like (Gini Index, Information, Entropy etc). Different type of splitting criterion is previously discussed in detailed in the decision tree model.

Step 4: Split the nodes into sub-nodes by using the best split.

Step 5: Keep repeating the above steps till the final leaf nodes have been obtained and T number of decision trees is built.

Step 6: Predict the new data by combining together the prediction of the n_{tree} trees. In classification the final output of the Random Forest is obtained based on the majority vote of the most popular class; and in regression estimate, the dependent variable is obtained by taking the average of their response.

Figure 5-5 displays the growing of Random Forest from multiple decision trees.

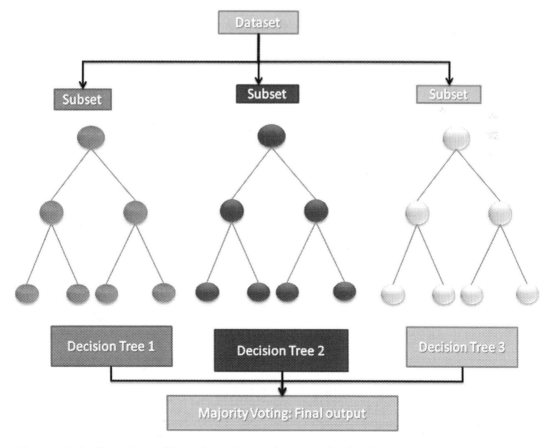

Figure 5-5. *Growing of Random Forest from multiple decision trees*

A Random Forest is an ensemble model where predictions from different decision trees are combined to obtain the final output or predictions. In classification problems, predictions of different trees could be combined by majority voting and in regression problems, it is obtained by taking the averages.[27] Here is the example taken from the healthcare industry to predict cancer remission in the patients by analyzing the patients' medical records. In this classification, problems can be seen in Figure 5-6 that how an ensemble model formed by majority voting produces more accurate predictions than the individual decision trees.

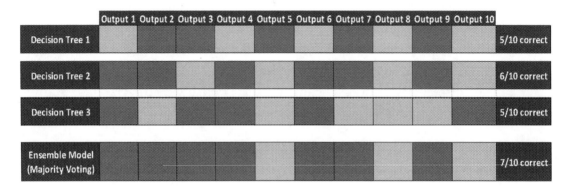

Figure 5-6. *Ensemble model formed by majority voting*

In the preceding Figure 5-6, all three decision tree individually predict 10 outputs in Green or Red color. In each individual decision tree, out of 10 outputs all correct prediction outputs are represented by Green, and incorrect predictions outputs are represented by Red. In decision tree 1 it is seen that there are 5 Green color outputs out of 10 outputs so the correct prediction is 5/10, in decision tree 2 there are 6 Green color outputs out of 10 outputs so the correct prediction is 6/10, similarly in decision tree 3 there are 5 Green outputs out of 10 outputs so the correct prediction is 5/10, and in the ensemble model there are 7 Green color outputs out of 10 outputs so the correct prediction is 7/10.

Hence the conclusion is that after aggregating all three decision trees based on the majority voting of correct and incorrect predictions, the ensemble model produces the highest predictive accuracy as compared to the individual decision tree. From the above example, it is seen that cancer remission in patients is accurately predicted (7/10 = correct prediction) using ensemble or Random Forests models.

The next example is to explain Random Forests in regression problems. The example is to predict hospital expenditures. In order to make the example simple, we are considering hospital expenditures as a dependent variable and age, length of stay, and gender as independent variables. In this example we will prepare three models. In this regression problem it can be seen in Figure 5-7 that the ensemble model is formed by averaging the predictions from the individual model.

Mentioned below are three expense brackets of Hospital Expenditure.

Hospital Expenditure:

Expense_bracket1: Less than $20k

Expense_bracket2: Between $20k - $60k

Expense_bracket3: Greater than $60k

Below are the outputs of the three different models based on Age, Gender, and Length_of_stay.

Model 1 Variable Age

	Hospital_expenditure	Expense_bracket1	Expense_bracket2	Expense_bracket3
	< 17	3%	40%	57%
Age	between 17 - 30	5%	35%	60%
	between 31 - 60	5%	35%	60%
	> 60	7%	28%	65%

Model 2 Variable Gender

	Hospital_expenditure	Expense_bracket1	Expense_bracket2	Expense_bracket3
Gender	Female	5%	35%	60%
	Male	7%	28%	65%

Model 3 Variable Length_of_stay

	Hospital_expenditure	Expense_bracket1	Expense_bracket2	Expense_bracket3
	0-1	5%	40%	55%
Length_of_stay	between 2 -7	6%	35%	59%
	> 7	5%	20%	75%

In order to form ensemble model Age >60, Gender is female and Length_of_stay >7 is considered. For each of these models below is the distribution of hospital expenditures across the Expense_bracket.

Ensemble Model

Model	Hospital_expenditure	Expense_bracket1	Expense_bracket2	Expense_bracket3
Age	> 60	7%	28%	65%
Gender	Male	7%	28%	65%
Length_of_stay	> 7	5%	20%	75%
Final output(Averaging)		6%	25%	68%

Figure 5-7. *Ensemble model formed by taking the average of hospital expense*

The final predictions of the hospital expenditures are simply calculated by taking the average of the hospital expenditure predictions in the same Expense_ bracket in a different model. For example, from the above analysis in Expense_bracket 3, taking the average of $\frac{65\% + 65\% + 75\%}{3} = \mathbf{68\%}$ and in Expense_bracket 2, taking the average of $\frac{28\% + 28\% + 20\%}{3} = \mathbf{25\%}$; and similarly for Expense_bracket 1 the average is **6%**. The conclusion from the above analysis is that there is **68%** of probability that final prediction of hospital expenditure is falling in Expense_bracket3 (greater than $60k), **25%** of probability that the final prediction of hospital expenditure is falling in Expense_bracket2 (between $20k – $60k), and **6%** of probability that the final prediction of hospital expenditure is falling in Expense_bracket1 (less than $20k).

There are other classification procedures like support vector machines (SVMS),[28] artificial neural networks (ANNS),[29] and boosted trees.[30] All these methods like Random Forests are highly accurate classifiers and can be applicable in both classification and regression problems. There are many numbers of advantages and disadvantages of Random Forests. Like all models, Random Forest has specific advantages as well as certain limitations. These are discussed in the following section.[31]

Advantages of Random Forests

- Random Forests is applicable to both classification and regression problems.

- Very high classification accuracy and stability.

- Large databases can be easily and efficiently handled by Random Forests.

- Helpful in variables selection from large databases where there are more numbers of variables than observations and produce variable importance measures for each predictor variable.

- Random Forests are non-parametric so there are no formal distributional assumptions.

- Handle missing values and outliers effectively. It can be done by imputation methods and the other approach is that it can be easily processed by the built-in procedure of recursive partitioning or surrogate splits.

- Helpful in modeling the complex interactions among the predictor variables.

- Useful in reducing overfitting problem.

Limitations of Random Forests

- Random Forests are not easy to interpret.

- They do not perform well when applied in regression problems.

- If the data consists of categorical variables with different number of levels, then the Random Forest algorithm gets biased in favour of those attributes that have more levels, hence variable importance score are not reliable in such type of data.[32] Methods such as partial permutations were used to solve the problem.

- If the data consists of correlated independent variables means there is multicollinearity in the data; in that case Random Forests variable importance measure is not reliable and can be misleading.

How to Select N$_{trees}$ in Random Forest?

Random Forest is a highly powerful and efficient machine learning technique that can operate quickly over huge data sets. Now the next question is how to determine the number of decision trees in composing a Random Forest. The general rule of thumb for deciding the optimal number of decision trees is based on an out-of-bag (OOB) error rate; it means building the trees until the error no longer decreases. If there is no concern regarding the computation times, then more numbers of trees will give better and reliable estimates from out-of-bag predictions. Looking at the OOB error rate, more numbers of trees are grown and stop once they level out, because a threshold is achieved after that, and increasing the number of trees would not bring any significant performance gain unless a huge computational environment is available.[33]

How to Select m_{try} in Random Forest?

The next important thing in building a Random Forest is to decide how many features or variables (m_{try}) must be considered for each split of the tree.

Recommended default in classification trees uses **sqrt P** and in regression trees it uses **P/3,** but sometimes it can be done based on trial and error and finding out where the model is having high predictive accuracy.[34]

Let's take an example of a healthcare training data set having 1,000 observations and 25 variables, where the number of features or variables for each split of the classification trees is calculated as **sqrt P,** which is **sqrt 25**; hence the number of features or variables used for each split is **5.** In a regression tree the number of features or variables for each regression tree is calculated as **P/3,** which is **25/3**; hence the number of features or variables used for each split is **8.**

Out-of-Bag (OOB) Error in Random Forests

In the Random Forests technique, there is no need for a separate validation set to get an unbiased estimate of the validation set error, the estimation is done internally during the run. In Random Forests each tree is built using a different bootstrap sample with replacement from the original data. In this sampling, about two-thirds of data is used for training and is called in-bag samples; and about one-third of the cases are left out of the bootstrap sample and can be used for validation and called out-of-bag samples, and the error estimated on these out-of-bag samples is the out-of-bag error.[35] The **Out-of-bag (OOB)** error or misclassification rate is a method for measuring the prediction error of the Random Forests. The Random Forest error rate depends on the correlation between any two trees in the Random Forests. If the correlation between the trees increases, the error rate also increases. If the correlation decreases, then the error rate also decreases and so does the strength of each individual tree in the Random Forests. If the strength of each individual tree increases, the error rate decreases; and if the strength of each individual trees; decreases, the error rate increases.

Variable Importance Measures in Random Forests

In Random Forest procedures, the two methods that help in estimating the important predictor variables in the model are **Mean Decrease Accuracy (MDA)** and **Mean Decrease Gini (MDG)**. Mean Decrease Accuracy (MDA) and Mean Decrease Gini (MDG) differ slightly from each other. Mean Decrease Accuracy (MDA) is also known as Permutation measure and is calculated as the normalized difference between the unpermuted observations of the test set of each variable and randomly permuted observations of the test set of each variable. In order to normalize the variable importance, each variable is divided by the standard error of each variable, which is taken from the parameter "importanceSD" from the model. A higher Mean Decrease Accuracy (MDA) value denotes a higher variable importance.[36] Mean Decrease Gini (MDG) is also known as **IncNodePurity** and is calculated by adding all of the decreases in Gini impurity at each tree node split normalized by the number of trees. Mean Decrease Gini (MDG) measures the purity of the variable. A higher IncNodePurity value denotes a higher variable importance (i.e., those nodes are pure).[37] When the number of variables are large, then a Random Forest is built by using all the variables; and then selecting the most important variables from the first run, the Random Forest model can be rebuilt again.

Proximity Measures in Random Forests

Proximity is defined as the nearness between pairs of cases or observations. In Random Forests, proximity is calculated for each pair of cases or observations; if two cases have the same terminal node through one tree, then their proximity is increased by one. Combine the overall trees in Random Forests and normalize by twice the number of trees in them.[38] The proximities formed the N*N matrix. The main diagonal cases in the proximity matrix will have perfect proximity to itself, that is, proximities close to one and it means that those observations or cases are similar while proximities close to zero mean that those observations or cases are dissimilar.[39] Proximities are used in missing values replacement for supervised and unsupervised data, locating outliers,[40] instructive data visualization using metric scaling,[41] and variable importance stratification.[42] In large datasets the proximity N*N matrix does not fit properly so in order to handle large datasets, the Random Forests use the "compressed" form of proximity matrix.

Random Forests Model Using R

In this healthcare case study, we will discuss the data and the variables used in the data. We then discuss about the exploratory data analysis in R, which is considered as the first step in the data analysis process. We also discuss about building a logistic regression model and interpretation of its output in R.

Business Problem: To predict the probability of malignant and benign breast cancer.

Business Solution: To build the Random Forest Model.

About Data

In this healthcare case study, use of the Random Forests model to predict the probability of malignant and benign breast cancer is demonstrated. The Random Forests model is created on synthetically generated data. Data was synthesized using Breast Cancer Wisconsin (Diagnostic) Data set that is available from the UCI Machine Learning Repository.[43] In this dataset there are total of **600 observations** and **11 variables**: 10 variables are numeric and 1 variable is categorical.

The synthetic_cancer_ data set contains 11 cytological characteristics of breast Fine- Needle-Aspirate (FNAs) that are assigned to a value between 1–10. The data set is of **600** patients' Fine-Needle-Aspirate(FNAs) cytological analysis. The outcome is the dependent or target variable in the data, where Yes denotes presence of malignant cancer and No denotes absence of malignant cancer. **Sixty-three** percent of observations are benign and **37** percent of observations are malignant. Variables in the data set, Thickness_of_Clump, Normal_Nucleoli Cell_Size_Uniformity, Cell_Shape_Uniformity, Marginal_Adhesion, Bland_Chromatin Single_Epithelial_Cell_Size, Bare_Nuclei, and Mitoses are assigned values on a scale of 1 to 10, with 1 being very close to benign and 10 being the most malignant.

In R, Create your own working directory to import the dataset.

```
#Read the data from the working directory, create your own working directly
to read the dataset.

data1 <- read.csv ("C:/Users/Deep/Desktop/data/
synthetic_cancer_data.csv",header=TRUE,sep=",")

data2<-data.frame(data1)
```

Performing Data Exploration

In Exploratory data analysis, we are going to take a broad look at patterns, trends, summary, outliers, missing values, and so on in the existing data. R codes for data exploration and their output is discussed in the section below.

```
#perform exploratory data analysis to know about the data.

# display top 6 rows of dataset to see how data look like

head (data2)
```

Sample_No	Thickness_of_Clump	Cell_Size_Uniformity	Cell_Shape_Uniformity
1	6	2	2
2	6	6	5
3	3	2	3
4	7	9	9
5	6	3	1
6	8	10	10

Marginal_Adhesion	Single_Epithelial_Cell_Size	Bare_Nuclei	Bland_Chromatin
2	2	2	5
6	7	10	5
1	2	2	4
2	4	4	4
4	3	3	4
9	7	10	9

Normal_Nucleoli	Mitoses	Outcome
1	1	No
2	2	No
2	2	No
7	2	No
3	1	No
9	2	Yes

```
# display bottom 6 rows to see how data look like
```

tail(data2)

Sample_No	Thickness_of_Clump	Cell_Size_Uniformity	Cell_Shape_Uniformity
595	4	8	7
596	7	3	1
597	6	2	4
598	6	1	3
599	5	3	2
600	6	4	5

Marginal_Adhesion	Single_Epithelial_Cell_Size	Bare_Nuclei	Bland_Chromatin
3	5	10	9
1	4	3	4
1	2	1	4
1	2	1	4
3	2	1	2
2	1	3	1

Normal_Nucleoli	Mitoses	Outcome
1	2	Yes
1	2	No
1	2	No
1	2	No
3	2	No
2	1	No

describe the structure of data, it displays the datatype of each variable
present in the data like whether that particular varibale is numeric,
factor etc .

str(data2)

```
'data.frame'              : 600 obs. of  11 variables:
 $ Sample_No             : int  1 2 3 4 5 6 7 8 9 10 ...
 $ Thickness_of_Clump    : int  6 6 3 7 6 8 3 4 4 6 ...
 $ Cell_Size_Uniformity  : int  2 6 2 9 3 10 2 3 2 2 ...
 $ Cell_Shape_Uniformity : int  2 5 3 9 1 10 1 2 1 3 ...
 $ Marginal_Adhesion     : int  2 6 1 2 4 9 1 2 2 1 ...
```

```
$ Single_Epithelial_Cell_Size: int  2 7 2 4 3 7 3 2 3 4 ...
$ Bare_Nuclei                 : int  2 10 2 4 3 10 10 1 3 1 ...
$ Bland_Chromatin             : int  5 5 4 4 4 9 5 3 1 2 ...
$ Normal_Nucleoli             : int  1 2 2 7 3 9 2 2 3 2 ...
$ Mitoses                     : int  1 2 2 2 1 2 1 2 5 1 ...
$ Outcome                     : Factor w/ 2 levels "No","Yes": 1 1 1 1 1 2
                                1 1 1 1 ...
```

```
# Remove Sample_No from data 2

data2 <- data2[,-1]

#display the column name of the data

names(data2)
[1] "Thickness_of_Clump" "Cell_Size_Uniformity" "Mitoses"

[4] "Cell_Shape_Uniformity" "Marginal_Adhesion" "Bland_Chromatin"

[7] "Single_Epithelial_Cell_Size" "Bare_Nuclei" "Normal_Nucleoli"

[10]"Outcome"

# display the datatype

class(data2)

[1] "data.frame"

# Check the missing values present in the data

sum(is.na(data2))
[1] 0

#to check the proportion of benign and malignant breast
cancer in the data

table(data2$Outcome)/nrow(data2)

    No        Yes
 0.6333333   0.3666667
```

The number represents that there are around 63% benign and 37% malignant breast cancer patients diagnosed.

#Model Building & Interpretation on Full Data

In this case, initially we are using complete data for model building and interpretation, and in the next section we are randomly splitting data into two parts: train data set and test data set. A detailed description is explained in the section below.

#install randomForest package

install.packages("randomForest")

library(randomForest)

To build Random Forests models, the RandomForest package need to be installed.

#Building Random Forests model on full data **Program1:**
```
rf_model <- randomForest(Outcome ~ ., data=data2, ntree=1500,mtry=3,importa
nce=TRUE)
```

rf_model

The prediction accuracy of the Random Forests of Program1 is summarized below:

```
Call:
 randomForest(formula = Outcome ~., data = data2,ntree = 1500,
 mtry = 3, importance = TRUE)
             Type of random forest: classification
                  Number of trees: 1500
     No. of variables tried at each split: 3

        OOB estimate of error rate: 3.67%

Confusion matrix:
         No    Yes    class.error
 No     366    14     0.03684211
 Yes    8      212    0.03636364
```

In above program1 randomForest function is used to build the model, Outcome is the dependent variable in the data, ntree = 1500 denotes the number of trees used in building the Random Forests model, by default ntree is 500, mtry = 3 denotes the number of random variables used at each split in building decision tree and `importance = True` denotes the important predictors used in the model. The OOB estimate denotes the predictive accuracy of the Random Forests model. The OOB estimate is also known as Out-of-Bag (OOB) error and in this case study OOB error or prediction error of Random Forests model is 3.67%.

`#plotting the rf_model`

`plot(rf_model)`

Error plot for rf_model of Program1 is displayed in Figure 5-8.

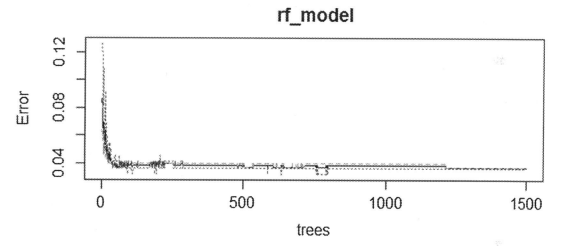

Figure 5-8. *Program1: Error plot for rf_model*

The error plot displayed in Figure 5-8 indicates the error for different classes (red and green color) and out-of-bag samples (black color) over the number of tress. Classes are in the same order as they are displayed in print (rf_model) so red = No and Green = Yes. From the above plot it is seen that around 1,200 the error is lowest and after that it levels out and an increase in the number of tresses will not provide any performance gain.

Splitting Data Set into Training and Testing

Now the data is split into two parts: train data set and test data set; the splitting ratio is 70:30, meaning that 70% of data contribute to the train dataset and 30% of data contribute to the test dataset.

```
#Set seed in order to reproduce the sample

set.seed(2)

#splitting data set into training and testing dataset in 70:30

install.packages("caTools")

library(caTools)

sample <- sample.split(data2$Outcome,SplitRatio=0.70)

#No of observations in train dataset

train_data <- subset(data2,sample==TRUE)

# No of observations  in test dataset

test_data <- subset(data2,sample==FALSE)
```

Synthetic_cancer_data set is divided into two – parts train_data and test_data, the splitting ratio is 70:30, meaning 70% of data contribute to the train_data set and 30% of data contribute to the test_data set. Train_data set is used to build the model and test_data set is used to test the performance of the model.

Model Building & Interpretation on Training and Testing Data

In this section the Train dataset is used to build the model and the test dataset is used to test the performance of the model.

```
#Building Random Forests model using training data
```

Program1.1:

```
r_model <- randomForest(Outcome ~ ., data=train_data,ntree=1500,mtry=3,
importance=TRUE)

print(r_model)
```

The prediction accuracy of the Random Forests of Program1.1 is summarized below:

```
Call:
 randomForest(formula = Outcome ~., data = train_data,
ntree = 1500,mtry = 3, importance = TRUE)

        Type of random forest: classification
             Number of trees: 1500
   No. of variables tried at each split: 3

     OOB estimate of error rate: 3.81%

Confusion matrix:
      No   Yes class.error
 No  256   10   0.03759398
 Yes  6   148   0.03896104
```

In the above Program1.1 code `randomForest`, `Outcome`, `ntree = 1500`, `mtry = 3` and `importance = True` is already discussed in the previous Program1 code.

OOB estimate denotes the predictive accuracy of the Random Forests model and in this case study OOB error or prediction error of Random Forests model is **3.81%.**

```
plot(r_model)
```

Error plot by number of trees for rf_model of Program1.1 is displayed in Figure 5-9.

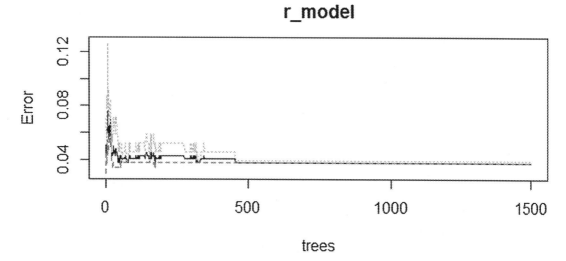

Figure 5-9. *Program1.1: Error plot by number of trees for rf_model*

Error plot by number of trees displayed in Figure 5-9 indicates the error for different classes (red and green color) and out-of-bag samples (black color) over the number of tresses. Classes are in the same order as it is displayed in print (rf_model) so red = No and Green = Yes. From the above plot it is seen that around 500 the error is lowest and after that the error levels out or flattens out and increase in number of tresses will not provide any performance gain.

```
plot(margin(r_model,test_data$Outcome))
```

Margin plot for r_model of Program1.1 is displayed in Figure 5-10.

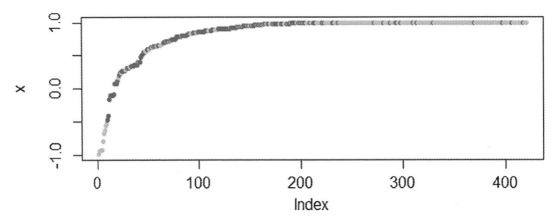

Figure 5-10. *Program1.1: Margin plot for rf_model*

In the Random Forests model, for classification problems the majority voting from all the decision trees is combined and then the most voted class is considered as the final predicted class. The margin is defined as the ratio of the votes for the correct class minus the maximum ratio of votes for other classes.[44] Margin function is the formula mentioned below:

$$mg(\mathbf{X},Y) = \frac{\sum_{k=1}^{K} I\big(h_k(\mathbf{X}) = Y\big)}{K} - \max_{j \neq Y} \left[\frac{\sum_{k=1}^{K} I\big(h_k(\mathbf{X}) = j\big)}{K} \right]$$

Where I(.) is indicator function.

$\mathbf{X},Y$ subscripts denote that the probability is over the $\mathbf{X},Y$ space.

If $mg(\mathbf{X},Y) > 0$ it means that the majority votes are correct and set of classifiers votes for the correct classification.

If $mg(\mathbf{X},Y) < 0$ it means that the majority votes are wrong and set of classifiers votes for the wrong classification.

Hence the larger is the value of margin, the more is the confidence in classification.

In Figure 5-10 it is seen that the classes are displayed by red and blue colors and a large proportion of the classes of No and Yes got classified correctly and a small proportion of classes in the ensemble model is not so sure or incorrectly classified.

```
# Predicting the model using test data

ran_pred <- predict(r_model,test_data)

#Display the confusion matrix or classification table

table(test_data$Outcome ,ran_pred)
```

In classification Table shown in Table 5-1 test_data$Outcome is considered as actual outcomes and ran_pred is considered as predicted outcomes.

Table 5-1. *Classification Table*

ran_pred		
test_data$Outcome	No	Yes
No	111	3
Yes	2	64

Diagonal values are correctly classified, hence Accuracy rate can be calculated as:

$$Accuracy\ rate = \frac{111+64}{111+3+2+64} = \frac{175}{180} = \mathbf{0.97}$$

Error rate or Misclassification error rate can be calculated as:

$$1 - Accuracy\ rate = 1 - 0.97 = \mathbf{0.03}$$

```
# Predicting the probability matrix  using test data

ran_prob <- predict(r_model,test_data,type = "prob" )

ran_prob

ran_prob1<-data.frame(ran_prob)
```

Probability matrix is displayed in Table 5-2.

Table 5-2. *Predicted Probability Table for Benign and Malignant Cancer*

Probability(No) Benign Cancer	Probability%(No)	Probability(Yes) Malignant Cancer	Probability%(Yes)
0.1200	12.00%	0.8800	**88.00%**
0.9987	99.87%	0.0013	0.13%
0.4753	47.53%	0.5247	52.47%
0.9080	90.80%	0.0920	9.20%
1.0000	100.00%	0.0000	0.00%
1.0000	100.00%	0.0000	0.00%
1.0000	100.00%	0.0000	0.00%
0.2600	26.00%	0.7400	**74.00%**
0.9993	99.93%	0.0007	0.07%
0.0380	3.80%	0.9620	**96.20%**

Probability (No) represents the probability value of presence of benign cancer in the patients and Probability% (No) represents the probability percentage of having benign cancer in the future and Probability (Yes) represents the probability value of the presence of malignant cancer in the patients and Probability% (Yes) represents the probability percentage of having malignant cancer in the future. For example, the probability percentage highlighted in bold font in Table 5-2 signifies that those patients are in a risky zone and it signifies that these three patients have **88.00%, 74.00%, and 96.20%** of probability to have malignant cancer in the future and need to be targeted as a priority with improved healthcare service and minimum cost; hence advance Predictive analytics helps in providing early warning signals in the healthcare industry and saving the life of thousands of patients.

Random Forests Model Using SAS

In this section, we discuss about different SAS procedures like proc content and proc freq. We also discuss about building a Random Forest model for predicting the probability of malignant and benign breast cancer on complete data, with explanation of SAS codes and output of each part in program 2, and in the next section we discuss about splitting data set into two parts – train data set and test data set – with an explanation of SAS codes and output of each part in program 2.1 and program 2.1.1 section.

```
/* Create your own library in SAS like here it is libref and mention the
path */
libname libref "/home/aro1260/deep";

/* Importing Synthetic_cancer_data */
PROC IMPORT DATAFILE= "/home/aroragaurav1260/data/synthetic_cancer_data.csv"
     DBMS=CSV Replace
     OUT=libref.cancer;
     GETNAMES=YES;
RUN;

/* To check the contents of the data */
PROC CONTENTS DATA=libref.cancer;
RUN;
```

In this case. partial output of procedure content is displayed in Table 5-3. It displays the content of the data like the number of observations; number of variables in the data; library name and data type of each variable; and whether the variables are numeric or character with their Length, Format, and Informat.

Table 5-3.

Part 1

The CONTENTS Procedure			
Data Set Name	LIBREF.CANCER	Observations	600
Member Type	DATA	Variables	11
Engine	V9	Indexes	0
Created	01/11/2018 15:41:09	Observation Length	88
Last Modified	01/11/2018 15:41:09	Deleted Observations	0
Protection		Compressed	NO
Data Set Type		Sorted	NO
Label			
Data Representation	SOLARIS_X86_64, LINUX_X86_64, ALPHA_TRU64, LINUX_IA64		
Encoding	utf-8 Unicode (UTF-8)		

Part 2

Alphabetic List of Variables and Attributes					
#	Variable	Type	Len	Format	Informat
7	Bare_Nuclei	Num	8	BEST12.	BEST32.
8	Bland_Chromatin	Num	8	BEST12.	BEST32.
4	Cell_Shape_Uniformity	Num	8	BEST12.	BEST32.
3	Cell_Size_Uniformity	Num	8	BEST12.	BEST32.
5	Marginal_Adhesion	Num	8	BEST12.	BEST32.
10	Mitoses	Num	8	BEST12.	BEST32.
9	Normal_Nucleoli	Num	8	BEST12.	BEST32.
11	Outcome	Char	3	$3.	$3.
1	Sample_No	Num	8	BEST12.	BEST32.
6	Single_Epithelial_Cell_Size	Num	8	BEST12.	BEST32.
2	Thickness_of_Clump	Num	8	BEST12.	BEST32.

```
/* Applying Proc freq to see the frequency of the data  */

proc freq data = libref.cancer;
tables Bare_Nuclei Outcome Outcome * Bare_Nuclei;
run;
```

Proc freq represents the number of frequency of each levels with cumulative frequency and cumulative percentage, like number of bare nuclei of different ranges from 1–10, number of benign breast cancer Patients (No) (63.33%), and number of malignant breast cancer patient's (Yes) (36.7%).

Table 5-4.

Part 1

The FREQ Procedure

Bare_Nuclei	Frequency	Percent	Cumulative Frequency	Cumulative Percent
1	124	20.67	124	20.67
2	117	19.50	241	40.17
3	122	20.33	363	60.50
4	26	4.33	389	64.83
5	22	3.67	411	68.50
6	14	2.33	425	70.83
7	12	2.00	437	72.83
8	12	2.00	449	74.83
9	12	2.00	461	76.83
10	139	23.17	600	100.00

Part 2

Outcome	Frequency	Percent	Cumulative Frequency	Cumulative Percent
No	380	63.33	380	63.33
Yes	220	36.67	600	100.00

Interaction between two variables like Outcome and Bare_Nuclei displays the table of outcome by Bare_Nuclei with frequency, percent, row percent, and column percent details. For example, the interaction between Outcome (No) and Bare Nuclei (1) the frequency is 120, percent is 20.00, row pct is 31.58, and col pct is 96.77.

Part 3

Frequency

Percent

Row Pct

Col Pct

Table of Outcome by Bare_Nuclei

Outcome	Bare_Nuclei 1	2	3	4	5	6	7	8	9	10	Total
No	120	111	111	17	8	5	2	1	1	4	380
	20.00	18.50	18.50	2.83	1.33	0.83	0.33	0.17	0.17	0.67	63.33
	31.58	29.21	29.21	4.47	2.11	1.32	0.53	0.26	0.26	1.05	
	96.77	94.87	90.98	65.38	36.36	35.71	16.67	8.33	8.33	2.88	
Yes	4	6	11	9	14	9	10	11	11	135	220
	0.67	1.00	1.83	1.50	2.33	1.50	1.67	1.83	1.83	22.50	36.67
	1.82	2.73	5.00	4.09	6.36	4.09	4.55	5.00	5.00	61.36	
	3.23	5.13	9.02	34.62	63.64	64.29	83.33	91.67	91.67	97.12	
Total	124	117	122	26	22	14	12	12	12	139	600
	20.67	19.50	20.33	4.33	3.67	2.33	2.00	2.00	2.00	23.17	100.00

```
/*Dropping the Sample_No from the data*/
data libref.final_cancer;
set libref.cancer(drop= Sample_No);
RUN;
PROC CONTENTS DATA=libref.final_cancer;
RUN;
```

Model Building & Interpretation on Full Data

In this section, we are using SAS Studio and complete data is used for model building and interpretation; and in the next section, we are randomly splitting data into two parts – training data set and testing data set.

In Program 2 code PROC HPFOREST procedure is used for building the Random Forests model, maxtrees indicates the number of trees used in the model, in this case study **maxtrees=1500**, vars_to_try indicates that 3 out of 9 input variables are randomly selected to be considered for a splitting rule, in this case study **vars_to_try =3** as in classification problem according to the rule it is ($\sqrt{9}$ = 3). Target variable is **Outcome** and it is categorical; therefore its level is **binary** and all **input** variables are numeric and their level is **nominal**. PROC HPFOREST runs on the final_cancer dataset and saves the model in a binary file Random_model_fit.bin, in this case **save file** option provides the full path for saving the model in a binary file **Random_model_fit.bin** and **ods output fitstatistics** option saves the fitstatistics output in **libref.fitstats_out**.

```
/*Building Random Forests model on full data */
```

/*Program2:*/

```
Proc hpforest data = libref.final_cancer maxtrees=1500 vars_to_try=3 ;
target Outcome/level=binary;
input  Thickness_of_Clump Cell_Size_Uniformity Cell_Shape_Uniformity
Marginal_Adhesion Single_Epithelial_Cell_Size Bare_Nuclei Bland_Chromatin
Normal_Nucleoli Mitoses/level=nominal;
ods output fitstatistics = libref.fitstats_out;
save file = "/home/aroragaurav1260/data/Random_model_fit.bin";
RUN ;
```

The **Random Forests output of program2** is split into several sections and each is discussed in the following section.

Parts 1, 2 of the Program2 output displays the information that the program ran on single machine (locally) and used 2 threads.

Part 1

The HPFOREST Procedure

Performance Information

Execution Mode	Single Machine
Number of Threads	2

Part 2

Data Access Information

Data	Engine	Role	Path
LIBREF.FINAL_CANCER	V9	Input	On Client

Part 3 of the Program2 output displays the information that few parameters are specified in the PROC HPFOREST statement; therefore maximum values are default except maxtrees=1500, which indicates the number of trees used in building Random Forests model and vars_to_try=3 ($\sqrt{9} = 3$) indicates that 3 out of 9 input variables are randomly selected to be considered for a splitting rule.

Part 3

Model Information

Parameter	Value	
Variables to Try	3	
Maximum Trees	1500	
Inbag Fraction	0.6	(Default)
Prune Fraction	0	(Default)
Prune Threshold	0.1	(Default)
Leaf Fraction	0.00001	(Default)
Leaf Size Setting	1	(Default)
Leaf Size Used	1	
Category Bins	30	(Default)
Interval Bins	100	

(continued)

Model Information

Parameter	Value	
Minimum Category Size	5	(Default)
Node Size	100000	(Default)
Maximum Depth	20	(Default)
Alpha	1	(Default)
Exhaustive	5000	(Default)
Rows of Sequence to Skip	5	(Default)
Split Criterion	.	Gini
Preselection Method	.	BinnedSearch
Missing Value Handling	.	Valid value

Part 4 of the Program2 output displays the information that the number of observations read and used is 600; it means there are no missing values present in the data.

Part 4

Number of Observations

Type	N
Number of Observations Read	600
Number of Observations Used	600

Part 5 of the Program2 output displays the baseline fit statistics. In PROC HPFOREST baseline fit statistics is computed first without using a model. The misclassification rate in the baseline fit statistics is **0.367** because that is the ratio of the observations for which it is malignant breast cancer (Yes).

Part 5

Baseline Fit Statistics

Statistic	Value
Average Square Error	0.232
Misclassification Rate	0.367
Log Loss	0.657

Part 6 of the Program2 output displays the first 10 and last 10 observations of the fit statistics output. In a single machine mode PROC HPFOREST computes fit statistic for the series of forests with increasing number of trees. The fit statistics improves or its value decreases with the increase in number of tresses initially and then after reaching to a certain stage, it levels off and fluctuates in a small range. Random Forests model provides the information of Average square error (Train), Average square error (OOB), Misclassification rate (Train), and Misclassification rate (OOB). Misclassification rate (OOB) is also known as out-of-bag (OOB) estimates and are less biased. From the above output the OOB misclassification rate of the model is fluctuating between **0.07** and **0.04** and is less than the misclassification rate in the baseline fit statistics of **0.367**, hence the model is considered a good model

Part 6 (top 10 observations)

Fit Statistics

Number of Trees	Number of Leaves	Average Square Error (Train)	Average Square Error (OOB)	Misclassification Rate (Train)	Misclassification Rate (OOB)	Log Loss (Train)	Log Loss (OOB)
1	15	0.0541	0.0739	0.0650	0.0750	0.498	0.979
2	29	0.0436	0.0826	0.0600	0.0974	0.207	1.125
3	50	0.0394	0.0823	0.0483	0.1000	0.173	0.938
4	68	0.0359	0.0752	0.0417	0.0932	0.167	0.809
5	84	0.0331	0.0671	0.0367	0.0860	0.160	0.645
6	98	0.0341	0.0616	0.0367	0.0870	0.130	0.405
7	114	0.0349	0.0606	0.0383	0.0872	0.131	0.366
8	130	0.0353	0.0554	0.0433	0.0780	0.132	0.250
9	145	0.0339	0.0523	0.0350	0.0688	0.129	0.179
10	158	0.0331	0.0497	0.0317	0.0637	0.127	0.170

Part 6 (last 10 observations)

Fit Statistics

Number of Trees	Number of Leaves	Average Square Error (Train)	Average Square Error (OOB)	Misclassification Rate (Train)	Misclassification Rate OOB)	Log Loss (Train)	Log Loss (OOB)
1490	24697	0.0305	0.0405	0.0350	0.0467	0.123	0.151
1491	24713	0.0305	0.0405	0.0350	0.0467	0.123	0.151
1492	24728	0.0305	0.0405	0.0350	0.0467	0.123	0.151
1493	24747	0.0305	0.0405	0.0350	0.0467	0.123	0.151
1494	24767	0.0305	0.0405	0.0350	0.0467	0.123	0.151
1495	24786	0.0305	0.0405	0.0350	0.0467	0.123	0.151
1496	24800	0.0305	0.0405	0.0350	0.0467	0.123	0.151
1497	24816	0.0305	0.0405	0.0350	0.0467	0.123	0.151
1498	24835	0.0305	0.0405	0.0350	0.0467	0.123	0.151
1499	24855	0.0305	0.0405	0.0350	0.0467	0.123	0.151
1500	24870	0.0305	0.0405	0.0350	0.0467	0.123	0.151

Part 7 of the Program2 output displays the information of Number of rules, which signifies the number of splitting rules that use a variable and loss reduction variable importance measures the variable importance two times – one time it is on training data and other time it is on OOB data. It can be seen from the fit statistics that OOB estimates are less biased. There are two measures OOB Gini and OOB Margin and OOB Gini is a more stringent measure and hence rows sorting is done based on OOB Gini meaasures. The OOB Gini is negative for two variables (Marginal_Adhesion and Mitoses) and OOB Margin is negative for one variable (Mitoses). The conclusion of fitting the Random Forests model to the data is that Bare_Nuclei, Cell_Size_Uniformity and Cell_Shape_Uniformity are the top three important predictors of the future onset of malignant breast cancer.

Part 7

Loss Reduction Variable Importance

Variable	Number of Rules	Gini	OOB Gini	Margin	OOB Margin
Bare_Nuclei	3265	0.112204	0.10209	0.224407	0.22382
Cell_Size_Uniformity	2897	0.101099	0.09367	0.202197	0.19867
Cell_Shape_Uniformity	3358	0.097528	0.08778	0.195056	0.18584
Normal_Nucleoli	2203	0.017954	0.01358	0.035908	0.03161
Bland_Chromatin	2499	0.014967	0.00929	0.029934	0.02418
Thickness_of_Clump	2720	0.013948	0.00792	0.027896	0.02477
Single_Epithelial_Cell_Size	1970	0.010975	0.00686	0.021950	0.01690
Marginal_Adhesion	2241	0.003202	-0.00086	0.006404	0.00108
Mitoses	2217	0.001974	-0.00099	0.003948	-0.00350

```
/* PLOTTING MISCLASSIFICATION RATE FOR TRAINING DATA */

proc sgplot data=libref.fitstats_out;
title "Misclassification Rate for Training Data";
series x=Ntrees y=MiscALL;
yaxis label='OOB Misclassification Rate';
run;
```

Figure 5-11 displays the plot of misclassification rate for training data.

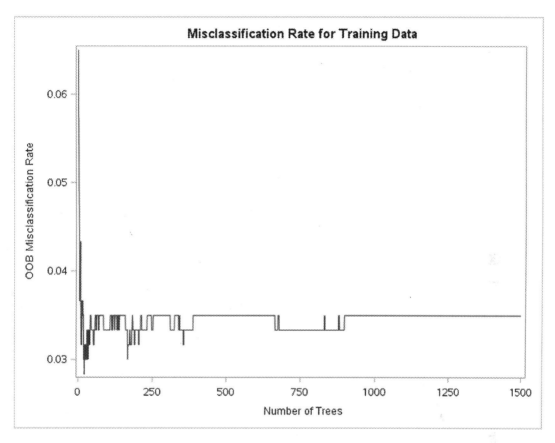

Figure 5-11. *Plot of misclassifictaion rate for training data*

Figure 5-11 shows that in training data the error is lowest or levels off around 1000.

```
/* Plot of OOB versus Training Misclassification Rate */

proc sgplot data=libref.fitstats_out;
title "OOB vs Training";
series x=Ntrees y=MiscAll;
series x=Ntrees y=MiscOob/lineattrs=(pattern=shortdash thickness=2);
yaxis label='Misclassification Rate';
run;
```

Figure 5-12 displays the plot of OOB versus Training Misclassification Rate.

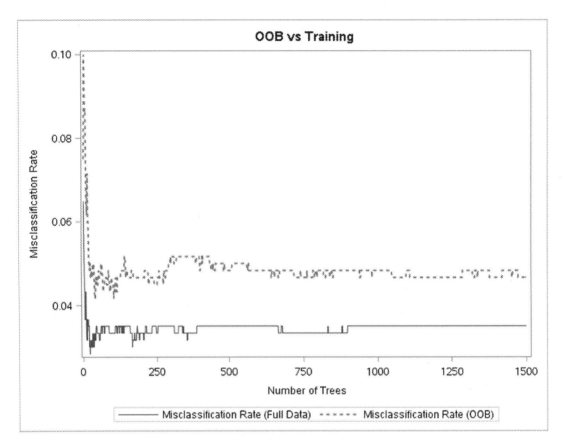

Figure 5-12. *Plot of OOB versus Training Misclassification Rate*

Figure 5-12 shows that the misclassification rate is larger based on the OOB data as compared to the misclassification rate based on training data, hence more numbers of trees are needed for the OOB misclassification rates to level off. From the plot it is clearly visible that in training data the error is lowest or level off around 1,000 whereas in OOB data it requires more numbers of trees to level off.

```
/*splitting data set into train and valid dataset in 70:30*/
```

In this section, we are randomly splitting data into 70 : 30 .

```
proc surveyselect data= libref.final_cancer method=srs seed=2 outall
samprate=0.7 out=libref.cancer_subset;
```

The SURVEYSELECT Procedure

Selection Method   Simple Random Sampling

The selection method used is simple random sampling.

Input Data Set	FINAL_CANCER
Random Number Seed	2
Sampling Rate	0.7
Sample Size	420
Selection Probability	0.7
Sampling Weight	0
Output Data Set	CANCER_SUBSET

Now divide the data into train and valid data sets based on the selection variable. Where selection is equal to 1, assign all those observations to train data set and where selection is equal to 0, assign all those observations to valid data set.

```
/*Values of selected variable: 1 means for train set, 0 means test set */

data libref.train;
   set libref.cancer_subset;
   if selected=1;

data libref.valid;
   set libref.cancer_subset;
   if selected=0;
```

Model Building & Interpretation on Training and Testing Data

The train data set is used to build the model and the valid data set is used to test the performance of the model.

```
data libref.train_valid;
set libref.cancer_subset;
Run;
```

/*Building Random Forests model using train data*/

/*Program2.1*/

```
proc hpforest data=libref.train_valid
maxtrees=1500 vars_to_try=3;
target Outcome/level=binary;
input Thickness_of_Clump
Cell_Size_Uniformity Cell_Shape_Uniformity
Marginal_Adhesion Single_Epithelial_Cell_Size
Bare_Nuclei Bland_Chromatin
Normal_Nucleoli Mitoses/level=nominal;
partition var= Selected (train = 1,valid = 0);
ods output VariableImportance= libref.loss_reduction_importance;
save file="/home/aroragaurav1260/data/Random_forest_fit.bin";
Run;
```

The **Random Forests output of program2**.1 is split into several sections and each of is discussed in the following section.

Parts 1, 2, and 3 of Program2.1 output is previously explained in Program2 output.

Part 1

The HPFOREST Procedure	
Performance Information	
Execution Mode	Single Machine
Number of Threads	2

Part 2

Data Access Information			
Data	**Engine**	**Role**	**Path**
LIBREF.TRAIN_VALID	V9	Input	On Client

Part 3

Model Information

Parameter	Value	
Variables to Try	3	
Maximum Trees	1500	
Inbag Fraction	0.6	(Default)
Prune Fraction	0	(Default)
Prune Threshold	0.1	(Default)
Leaf Fraction	0.00001	(Default)
Leaf Size Setting	1	(Default)
Leaf Size Used	1	
Category Bins	30	(Default)
Interval Bins	100	
Minimum Category Size	5	(Default)
Node Size	100000	(Default)
Maximum Depth	20	(Default)
Alpha	1	(Default)
Exhaustive	5000	(Default)
Rows of Sequence to Skip	5	(Default)
Split Criterion	.	Gini
Preselection Method	.	BinnedSearch
Missing Value Handling	.	Valid value

Part 4 of Program2.1 displays the number of observations read and used for train (420) and valid(180).

Part 4

Number of Observations

Type	NTrain	NValid	NTotal
Number of Observations Read	420	180	600
Number of Observations Used	420	180	600

Part 4 of Program2.1 displays the number of observations read and used for train (420) and valid(180).

Part 5 of the Program2.1 output displays the baseline fit statistics. In PROC HPFOREST baseline fit statistics is computed first without using a model. The misclassification rate in the baseline fit statistics is **0.371** because that is the ratio of the observations for which it is malignant breast cancer (Yes).

Part 5

Baseline Fit Statistics

Statistic	Value	Validation
Average Square Error	0.233	0.229
Misclassification Rate	0.371	0.356
Log Loss	0.660	0.651

Part 6 of the Program2.1 output displays the first 10 and last 10 observations of the fit statistics output. In a single machine mode PROC HPFOREST computes fit statistic for the series of forests with an increasing number of trees. The fit statistics improves or its value decreases with the increase in number of tresses initially and then after reaching to a certain stage it levels off and fluctuates in a small range. Random Forests model provides the information of Average square error (Train), Average square error (OOB), Average square error (Valid), and Misclassification rate (Train), Misclassification rate (OOB), and Misclassification rate (Valid). From the above output the Misclassification rate (Valid) is fluctuating between **0.09** and **0.07** and is less than the misclassification rate in the baseline fit statistics of **0.371**, hence the model is considered a good model.

Part 6 (top 10 observations)

Fit Statistics

Number of Trees	Number of Leaves	Average Square Error (Train)	Average Square Error (OOB)	Average Square Error (Valid)	Misclassification Rate (Train)	Misclassification Rate (OOB)	Misclassification Rate (Valid)	Log Loss (Train)	Log Loss (OOB)	Log Loss (Valid)
1	14	0.0761	0.0983	0.1060	0.0952	0.1071	0.1167	0.758	1.501	1.545
2	22	0.0503	0.0835	0.0713	0.0833	0.1161	0.1111	0.209	1.003	0.550
3	34	0.0486	0.0785	0.0704	0.0690	0.1090	0.0778	0.160	0.740	0.557
4	45	0.0388	0.0665	0.0619	0.0476	0.0836	0.0833	0.139	0.601	0.309
5	61	0.0369	0.0591	0.0573	0.0381	0.0610	0.0944	0.140	0.461	0.187
6	69	0.0357	0.0593	0.0595	0.0333	0.0692	0.0889	0.137	0.309	0.193
7	83	0.0355	0.0521	0.0586	0.0381	0.0569	0.0833	0.137	0.184	0.191
8	97	0.0360	0.0506	0.0569	0.0405	0.0636	0.0833	0.138	0.179	0.189
9	110	0.0349	0.0493	0.0585	0.0381	0.0680	0.0833	0.136	0.171	0.197
10	123	0.0343	0.0494	0.0600	0.0357	0.0673	0.0833	0.135	0.173	0.202

Part 6 (last 10 observations)

Fit Statistics

Number of Trees	Number of Leaves	Average Square Error (Train)	Average Square Error (OOB)	Average Square Error (Valid)	Misclassification Rate Train)	Misclassification Rate (OOB)	Misclassification Rate (Valid)	Log Loss (Train)	Log Loss (OOB)	Log Loss (Valid)
1490	18607	0.0307	0.0412	0.0507	0.0310	0.0500	0.0667	0.129	0.158	0.181
1491	18620	0.0307	0.0412	0.0506	0.0310	0.0500	0.0667	0.129	0.158	0.181
1492	18634	0.0307	0.0412	0.0506	0.0310	0.0500	0.0667	0.129	0.158	0.181
1493	18646	0.0307	0.0412	0.0506	0.0310	0.0500	0.0667	0.129	0.158	0.181
1494	18658	0.0307	0.0412	0.0506	0.0310	0.0500	0.0667	0.129	0.158	0.181
1495	18668	0.0307	0.0412	0.0506	0.0310	0.0500	0.0667	0.129	0.158	0.181
1496	18677	0.0307	0.0411	0.0506	0.0310	0.0500	0.0667	0.129	0.158	0.181
1497	18688	0.0307	0.0412	0.0506	0.0310	0.0500	0.0667	0.129	0.158	0.181
1498	18700	0.0307	0.0412	0.0506	0.0310	0.0500	0.0667	0.129	0.158	0.181
1499	18711	0.0307	0.0412	0.0506	0.0310	0.0500	0.0667	0.129	0.158	0.181
1500	18723	0.0307	0.0412	0.0506	0.0310	0.0500	0.0667	0.129	0.158	0.181

Part 7 of the Program2.1 output is previously discussed in Program2 output.

Part 7

Loss Reduction Variable Importance

Variable	Number of Rules	Gini	OOB Gini	Valid Gini	Margin	OOB Margin	Valid Margin
Cell_Shape_ Uniformity	2485	0.106541	0.09673	0.08171	0.213082	0.20463	0.18548
Bare_Nuclei	2172	0.111845	0.09588	0.08347	0.223691	0.22511	0.21172
Cell_Size_ Uniformity	2236	0.089356	0.08036	0.08199	0.178713	0.17278	0.17663
Normal_ Nucleoli	1649	0.020824	0.01518	0.01468	0.041649	0.03710	0.03548
Single_ Epithelial_ Cell_Size	1619	0.012357	0.00858	0.00840	0.024715	0.01722	0.01737
Bland_ Chromatin	1838	0.012302	0.00641	0.00750	0.024604	0.01864	0.01886
Thickness_of_ Clump	1849	0.010342	0.00578	0.00654	0.020684	0.01614	0.01940
Mitoses	1743	0.002927	-0.00013	-0.00186	0.005853	-0.00083	-0.00295
Marginal_ Adhesion	1632	0.002765	-0.00084	-0.00070	0.005530	-0.00111	-0.00048

In Program 2.1.1 code HP4SCORE procedure is used after HPFOREST procedure to create a model and save the model in a binary model file. PROC HP4SCORE applies the binary model file for the scoring purpose or estimating the variable importance in the particular data set. The IMPORTANT statement in validation data set is used in order to compute the importance of the variable based on the random branch assignment method (RBA). In PERFORMANCE statement THREADS =1 option is used so when the code is run again and again, there is not any change made in random branch assignments. IMPORTANT FILE is used in order to refer the file or the full path of the file that is created by PROC HPFOREST. OUT is used to create the output dataset, and in this case output dataset is created and stored in library libref .scored (out = libref.scored), VAR is used in order to specify all the input variables so their variable importance can be computed.

```
/* Predicting the model using valid data */
```

/*Program2.1.1 */

```
proc hp4score data= libref.valid;
ods output VariableImportance=libref.rba_importance_valid;
performance threads=1;
importance file=
"/home/aroragaurav1260/data/Random_forest_fit.bin"
out=libref.scored
var=(Outcome Thickness_of_Clump Cell_Size_Uniformity Cell_Shape_Uniformity
Marginal_Adhesion Single_Epithelial_Cell_Size Bare_Nuclei Bland_Chromatin
Normal_Nucleoli Mitoses);
Run;
```

The **Random Forests output of program2.1.1** for predicting on **valid data** is split into several sections and each of is discussed in the following section.

Parts 1, 2, and 3 of the Program2.1.1 output displays the information that the program ran on a single machine (locally) and used 1 thread and number of observations read and used in valid data is 180.

Part 1

The HP4SCORE Procedure

Performance Information

Execution Mode	Single Machine
Number of Threads	1

Part 2

Data Access Information

Data	Engine	Role	Path
LIBREF.VALID	V9	Input	On Client
LIBREF.SCORED	V9	Output	On Client

Part 3

Number of Observations	
Type	N
Number of Observations Read	180
Number of Observations Used	180
Sum of Frequencies Used	180

Part 4 of the program2.1.1 displays the information of the variable importance based on random branch assignments (RBA). Random branch assignments are less biased than loss reduction and less influenced by correlations. RBA uses the Margin and MSE.

Part 4

Random Branch Assignments Variable Importance		
Variable	Margin	MSE
Cell_Size_Uniformity	0.15906	0.04216
Bare_Nuclei	0.17286	0.03901
Cell_Shape_Uniformity	0.16009	0.03554
Normal_Nucleoli	0.03052	0.00667
Thickness_of_Clump	0.01798	0.00547
Bland_Chromatin	0.01421	0.00284
Single_Epithelial_Cell_Size	0.01333	0.00139
Marginal_Adhesion	0.00380	0.00019
Outcome	0.00000	0.00000
Mitoses	-0.00045	-0.00076

PROC SORT is used to sort the rba_importance_valid output file and the sorting is done based on MARGIN in DESCENDING order.

```
/*Sorting the rba_importance_valid by Margin /*

proc sort data = libref.rba_importance_valid;
by descending Margin;
Run;
```

```
proc print data= libref.rba_importance_valid;
Run;
```

Obs	Variable	MSE	Margin
1	Bare_Nuclei	0.03901	0.17286
2	Cell_Shape_Uniformity	0.03554	0.16009
3	Cell_Size_Uniformity	0.04216	0.15906
4	Normal_Nucleoli	0.00667	0.03052
5	Thickness_of_Clump	0.00547	0.01798
6	Bland_Chromatin	0.00284	0.01421
7	Single_Epithelial_Cell_Size	0.00139	0.01333
8	Marginal_Adhesion	0.00019	0.00380
9	Outcome	0.00000	0.00000
10	Mitoses	-0.00076	-0.00045

```
/*computing misclassification rate */

data libref.final_score;
set libref.scored ;
if upcase(Outcome) ne upcase(I_Outcome) then misclass=1;
else misclass=0;
run;
```

The MEANS procedure table displays the information of the misclassification error; in this case the misclassification error is 0.07.

```
proc means data=libref.final_score(where=(Outcome ne ''));
var misclass;
run;
```

The MEANS Procedure

Analysis Variable : misclass

N	Mean	Std Dev	Minimum	Maximum
1980	0.0702020	0.2555517	0	1.0000000

The conclusion of fitting the Random Forests model to the data is that **Bare_Nuclei, Cell_Shape_Uniformity,** and **Cell_Size_Uniformity** are the top three important predictors of the future onset of malignant breast cancer.

Summary

In this chapter we learned about the application of data analytics in the healthcare sector The model discussed in the chapter is the Random Forest model. Various characteristics, features, advantages, and limitations of the model were discussed. Practical application of this model was demonstrated by help of a case study to predict the probability of malignant and benign breast cancer. From the stage of Model development to execution and result, interpretations were performed and discussed in R and SAS Studio.

References

1. Deloitte 2018 Global health care outlook:The evolution of smart health care; Deloitte: 2018; p 32.

2. Langer, E. 2017 Biopharmaceutical Trends — Opportunities For The New Year Bioprocess Online [Online], 2016.

3. Administration, I. T. 2016 ITA Medical Devices Top Markets Report; Department of Commerce, USA: `https://www.trade.gov/topmarkets/pdf/Medical_Devices_Executive_Summary.pdf`

4. Patrick Finn, F. S., Tobias Schneider, and; Singhal, S., The growth opportunity for private health-insurance companies. McKinsey & Company: `https://www.mckinsey.com/industries/healthcare-systems-and-services/our-insights/the-growth-opportunity-for-private-health-insurance-companies`, 2017.

5. Aura-Maria Cardona, Z. S. R., Design Considerations in High-Throughput Automation for Biotechnology. In Florida Conference on Recent Advances in Robotics, FCRAR, Melbourne, Florida,USA, 2008; p 11.

6.   Lodha, A., Clinical Analytics – Transforming Clinical Development through Big Data. 2016 2016, 2 (10).

7.   Buszewski, B.; Kesy, M.; Ligor, T.; Amann, A., Human exhaled air analytics: biomarkers of diseases. Biomed Chromatogr 2007, 21 (6), 553-66.

8.   Neuman, M. R.; Baura, G. D.; Meldrum, S.; Soykan, O.; Valentinuzzi, M. E.; Leder, R. S.; Micera, S.; Zhang, Y. T., Advances in Medical Devices and Medical Electronics. Proceedings of the IEEE 2012, 100 (Special Centennial Issue), 1537-1550.

9.   Haydon, I. Biologics: The Drugs Transforming Medicine Healthcare of tomorrow [Online], 2017. https://www.usnews.com/news/healthcare-of-tomorrow/articles/2017-07-25/biologics-the-drugs-that-are-transforming-medicine.

10. Raghupathi, W.; Raghupathi, V., Big data analytics in healthcare: promise and potential. Health Inf Sci Syst 2014, 2, 3.

11. Cleverley, W. O., The health care industry: in evolution or revolution? J Health Care Finance 1999, 25 (4), 2-14.

12. association, B. c. b. s. Shopping for surgery: How consumers are saving with shift to outpatient care 2016, p. 10.

13. Morchel, H.; Raheem, M.; Stevens, L.; Horahan, K., Electronic health records access during a disaster. Online J Public Health Inform 2014, 5 (3), 232.

14. Smith, E.; Macdonald, R., Managing health information during disasters. Health Inf Manag 2006, 35 (2), 8-13.

15. O'Connor, S. Pros and Cons of Electronic Health Records 2017.

16. Meece, J., Effect of insulin pen devices on the management of diabetes mellitus. Am J Health Syst Pharm 2008, 65 (11), 1076-82.

17. Chew, C.; Eysenbach, G., Pandemics in the Age of Twitter: Content Analysis of Tweets during the 2009 H1N1 Outbreak. PLOS ONE 2010, 5 (11), e14118.

18. Echevarría-Zuno, S.; Mejía-Aranguré, J. M.; Mar-Obeso, A. J.; Grajales-Muñiz, C.; Robles-Pérez, E.; González-León, M.; Ortega-Alvarez, M. C.; Gonzalez-Bonilla, C.; Rascón-Pacheco, R. A.; Borja-Aburto, V. H., Infection and death from influenza A H1N1 virus in Mexico: a retrospective analysis. The Lancet 2009, 374 (9707), 2072-2079.

19. Jamei, M.; Nisnevich, A.; Wetchler, E.; Sudat, S.; Liu, E., Predicting all-cause risk of 30-day hospital readmission using artificial neural networks. PLoS One 2017, 12 (7), e0181173.

20. Joudaki, H.; Rashidian, A.; Minaei-Bidgoli, B.; Mahmoodi, M.; Geraili, B.; Nasiri, M.; Arab, M., Using data mining to detect health care fraud and abuse: a review of literature. Glob J Health Sci 2014, 7 (1), 194-202.

21. Tie, R., The Use of Integrated Methods to Fight Fraud. Fraud Magazine 2011, p 7.

22. Lobosco, K. Doctors and nurses busted for $712 million Medicare fraud 2015.

23. Ruiz, R. R. U.S. Charges 412, Including Doctors, in $1.3 Billion Health Fraud 2017.

24. Blumenthal, D.; Abrams, M.; Nuzum, R., The Affordable Care Act at 5 Years. New England Journal of Medicine 2015, 372 (25), 2451-2458.

25. Sorrell, J. M., Ethics: the Patient Protection and Affordable Care Act: ethical perspectives in 21st century health care. Online J Issues Nurs 2012, 18 (1).

26. Breiman, L., Random Forests. Machine Learning 2001, 45 (1), 5-32.

27. Kenneth Soo, A. N. Random Forests: Predicting crime in San Francisco Algobeans Layman tutorials in Analytics [Online], 2016.

28. Vishwanathan, S. V. M.; Murty, M. N. In SSVM: a simple SVM algorithm, Neural Networks, 2002. IJCNN '02. Proceedings of the 2002 International Joint Conference on, 2002; 2002; pp 2393-2398.

29. Reby, D.; Lek, S.; Dimopoulos, I.; Joachim, J.; Lauga, J.; Aulagnier, S., Artificial neural networks as a classification method in the behavioural sciences. Behavioural Processes 1997, 40 (1), 35-43.

30. Elith, J.; Leathwick, J. R.; Hastie, T., A working guide to boosted regression trees. J Anim Ecol 2008, 77 (4), 802-13.

31. Cutler, D. R.; Edwards, T. C.; Beard, K. H.; Cutler, A.; Hess, K. T.; Gibson, J.; Lawler, J. J., Random Forests for Classification in Ecology. Ecology 2007, 88 (11), 2783-2792.

32. Strobl, C.; Boulesteix, A.-L.; Zeileis, A.; Hothorn, T., Bias in random forest variable importance measures: Illustrations, sources and a solution. BMC Bioinformatics 2007, 8 (1), 25.

33. Mayumi Oshiro, T.; Santoro Perez, P.; Baranauskas, J., How Many Trees in a Random Forest? 2012; Vol. 7376.

34. Genuer, R.; Poggi, J.-M.; Tuleau-Malot, C., Variable selection using random forests. Pattern Recognition Letters 2010, 31 (14), 2225-2236.

35. Nguyen, T.-T.; Huang, J. Z.; Nguyen, T. T., Unbiased Feature Selection in Learning Random Forests for High-Dimensional Data. The Scientific World Journal 2015, 2015, 18.

36. Kuhn, S.; Egert, B.; Neumann, S.; Steinbeck, C., Building blocks for automated elucidation of metabolites: Machine learning methods for NMR prediction. 2008; Vol. 9, p 400.

37. Calle, M. L.; Urrea, V., Letter to the Editor: Stability of Random Forest importance measures. Briefings in Bioinformatics 2011, 12 (1), 86-89.

38. Ishioka, T. In Imputation of Missing Values for Unsupervised Data Using the Proximity in Random Forests, The Fifth International Conference on Mobile, Hybrid, and On-line Learning, 2013.

39. Louppe, G. Understanding random Forests from theory to practice. University of Liège, https://arxiv.org/pdf/1407.7502.pdf, 2014.

40. Tsuji, S.; Midorikawa, Y.; Takahashi, T.; Yagi, K.; Takayama, T.; Yoshida, K.; Sugiyama, Y.; Aburatani, H., Potential responders to FOLFOX therapy for colorectal cancer by Random Forests analysis. Br J Cancer 2012, 106 (1), 126-32.

41. Buja, A.; Swayne, D. F.; Littman, M. L.; Dean, N.; Hofmann, H.; Chen, L., Data Visualization With Multidimensional Scaling. Journal of Computational and Graphical Statistics 2008, 17 (2), 444-472.

42. Seoane, J.; Day, I.; Campbell, C.; Casas, J.; Gaunt, T. In Using a Random Forest proximity measure for variable importance stratification in genotypic data, 2nd International Work-Conference on Bioinformatics and Biomedical Engineering (IWBBIO), Granada, Spain, Ortuno, F. a. R., I, Ed. Granada, Spain, 2014.

43. Wolberg, W. H.; Mangasarian, O. L., Multisurface method of pattern separation for medical diagnosis applied to breast cytology. Proceedings of the National Academy of Sciences 1990, 87 (23), 9193-9196.

44. Koulis, T., Random Forests: Presentation Summary. University of Waterloo: http://www.math.uwaterloo.ca/~hachipma/stat946/koulis.pdf, 2003; p 11.

CHAPTER 6

Airline Case Study

The global airline industry provides air transport services for traveling passengers and freight to every corner of the world, and it has been an essential part of the formation of a global economy. The airline industry is a fast growing and dynamic sector. The International Air Transport Associations (IATA) forecasts that the global net profit to rise to $38.4 billion in 2018.[1] The airline industry is achieving sustainable profitability levels and 2018 is expected to be the fourth consecutive year of sustainable profits. According to IATA, in 2016 the aircraft carries nearly 3.8 billion passengers and expects 7.2 billion passengers to travel in 2035.[2] Boeing and Airbus are considered the world's biggest aircraft manufacturer.

There are over 3,500 air carriers globally with over 100 in the United States alone. According to June 2016 statistics, American Airlines is the world's largest airlines by aircraft fleet size, second is Delta Air Lines, and third is United Airlines;[3] all three are US companies.

The US Department of Transportation (DOT) classifies the air carriers based on revenue. Classified under three categories and revenue ranges, these carriers are displayed in Figure 6-1.

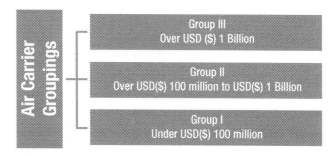

Figure 6-1. *Air carrier Groupings*

© Deepti Gupta 2018
D. Gupta, *Applied Analytics through Case Studies Using SAS and R*,
https://doi.org/10.1007/978-1-4842-3525-6_6

The World Airline Awards were conducted at the Paris Air Show on June 20, 2017.[4] According to Skytrax the prestigious award ceremony is known as the **"Oscars of the aviation industry"** and Qatar Airways has been considered the best airline in the world for 2017. The other top nine airlines in 2017, according to Skytrax, are Singapore Airlines, ANA All Nippon Airways, Emirates, Cathay Pacific, Eva Air, Lufthansa, Etihad Airways, Hainan Airlines, and Garuda Indonesia.[5]

The Airline Industry is growing at an incredible rate but is also facing certain challenges as it heads into the future. The three major challenges faced by the industry are discussed in the following section.

1. **Safety and Security**

    The biggest challenge faced by the airline industry is safety and security of passengers as well as their crew and assets. Global terrorism continues to be prevalent and the airline industry has historically been a target; one example is the Brussels attacks.[6] Government, airport authority, and airlines need to figure out the balance between an increased level of security and comfort to the passengers. Other two recent tragic examples are the Malaysia plane MH370 crash[7] and MH17 Ukraine plane crash.[8] The IATA has invested to build the data management platform known as Global Aviation Data Management (GADM).[9] The objective of the GADM program is to collect the data from the different sources like flight operations, Infrastructure, and IATA audits and then analyze the data by using advance analytical methods, which will improve the safety measures in the airline industry.

2. **Passenger Experience**

    Another big challenge faced by the airline industry is passenger experience. Over the years the travel experience has improved in certain areas but balancing the experience and increasing costs can require out-of-the-box thinking. A passenger's experience road map includes the improved experience from the time of the flight booking, transport, experience in airport premises, on-board experience, and post-flight experience. In today's world, airports are not only just the point of departure or arrival for the passengers but they are the center of leisure facilities. Technology is playing an essential role in making the boarding process easier and faster, providing the passengers with a better experience and more leisure time. The main focus is how to better serve the passengers by providing services like Wi-Fi, food and beverages, casinos, and beauty salons. These amenities make the passengers feel comfortable and also provide a venue for additional airport revenue. According to IATA, around 7.2 billion passengers are expected to travel in 2035. This indicates that with the increase in the passenger traffic and investment in infrastructure, the airports will likely to become bigger and more modernized.

3. **Sustainability**

Like all sectors, aviation is also taking various measures to reduce carbon emissions and adopt more environmental friendly practices. The third big challenge for the airline industry is how to modernize while decreasing the environmental impact. In order to achieve the goal of becoming carbon neutral and reduce the carbon emissions, **Airport Carbon Accreditation** was developed and launched by Airports Council International (ACI) Europe in 2009.[10] The program is contributing to achieve the overall sustainability goals of the airline industry as defined by ICAO and ATAG. All the airports throughout the world are encouraged to participate in an accreditation program and implement best practices in carbon management, resulting in reduction of carbon emissions. As of September 2017 there are total 198 airports that have been accredited to the program and a total of 34 airports worldwide have become carbon neutral including 27 in Europe, 5 in Asia-Pacific, 1 in North America, and 1 in Africa. The long-term expectation of the program is that by 2030 there must be 100 carbon neutral airports in Europe. Figure 6-2 displays the Four Ascending Levels of Certification to carbon neutrality.[11]

In this chapter the key applications of analytics in the airline industry is discussed. We then discuss the multiple linear regression model and the concepts and vocabularies used in regression analysis. We also provide a detailed case study on predicting the flight arrival delays in minutes by using R and SAS Studio.

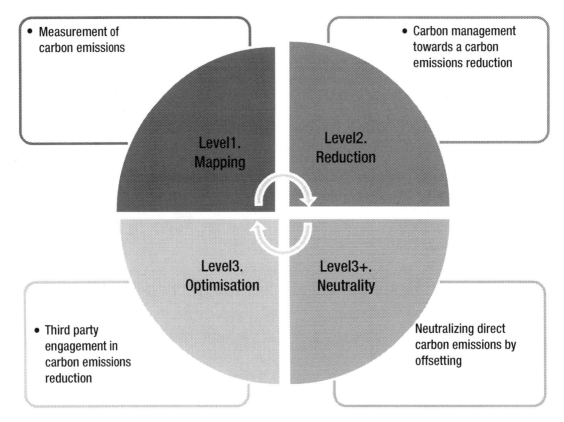

Figure 6-2. *Four Ascending Levels of Certification to Carbon Neutrality*

Application of Analytics in the Airline Industry

The airline industry produces an enormous amount of data on a daily basis. With the help of predictive modeling, the hidden patterns in the data can be revealed and effective insights can be drawn from the data. Advance predictive analytical techniques unlock a realm of opportunities for airport, airlines, travel agencies, and passengers. Data plays a major role in the airline industry from a customer loyalty program to fraud detection, flight safety, and predicting flight delays. Airline data is collected from different source like passenger traveling information, sensor data from the plane, etc. In today's time, the airline industry must consider them as the data companies first and travel companies second.

The airline industry is facing a big challenge in terms of managing the data produced on a daily basis in different formats like text, images, etc., by using traditional methods due to the shear high volume and variety in the data. In some cases, the carriers do not have proper IT infrastructure to capture, store, and analyze the enormous amount of the data generated on daily basis and some carriers does not have proper technological infrastructure, which will be a barrier for those carriers to use their data effectively. Big data technologies are being implemented for storage, accessibility, security, and overall management of this vast amount of data. This vast amount of data offers an opportunity to employ predictive analytics, machine learning, and Artificial Intelligence algorithms like Regression, Bayesian Network, artificial neural network, and SVM, etc., to effectively withdraw insights and translate it into usable information. Predictive Analytics is playing an important role in redefining the airline industry in a holistic sense with applications ranging from personalized offers and passenger experience, safer flights, airline fraud detection, and predicting flight delays. An overview of some of these key applications is provided in this section.

Personalized Offers and Passenger Experience

Today's passengers have a various number of choices and options available when they would like to book their airline tickets, which is creating a pressure for all of the airline industry to provide an outstanding passenger experience. A few years back a one-size-fits-all solutions strategy used to be implemented by the airline industry but due to digitization, advance technology, and online marketing, there is a drastic difference seen in the industry.[12] In today's time different passengers have different needs and priorities when it comes to air travel. Airline industries are using the better data to provide highly personalized tailor-made offers to individual travelers based on their needs, increasing the passengers' satisfaction and loyalty. Airline companies can easily collect the passenger data just from a single booking and make use of that data for generating better insights and knowing their passenger in a better way.

Passengers can be segmented based on their purpose of trip like business trip or leisure trip; and based on this, the behavior of the passengers can be predicted like whether the passenger is price sensitive or looking for product quality. Predictive analytics and machine learning algorithms like supervised classification, recommendation engine, clustering, sentiment analysis, etc., is applied in huge airline data. Cluster algorithms can be applied in order to segment the passengers based on their purchasing history, behaviors, and class.[13] Sentiment analysis can be done by monitoring the reviews and passengers' experience shared in social media platforms by the passengers during their traveling period. By extracting the tweets and analyzing the sentiment of the passengers like positive, negative' and neutral, these sentiments help the traveling companies to know about their products and services better and helps in improving the passenger's experience.[14] With the help of travel recommender systems prediction and the suggestion of products and services recommended to the passengers, like what the passengers want (their preferences, interests, locations, etc.) and target them with the tailor-made offers accordingly.[15] For example, bundle packages can be offered to the passengers with the flight, extra baggage, hotel at a better price, and rental cars. Travel recommender systems provide the best and relevant option at better prices to the passengers and helps in increasing the revenue for travel providers.

Safer Flights

The airline industry generates a huge amount of data related to the aircraft and flights. During a flight, the data is generated every second in a form of pilot report, incident report, control positions, and warning reports, etc. The data is collected for hundreds of sensors every second for each flight. Analyzing such an enormous amount of data manually is not feasible. Using Supervised and Unsupervised machine learning algorithms and predictive analytics, this enormous amount of data can be easily monitored and analyzed wisely, hence improving the aircraft safety. When out-of-trend data is observed, the automatic alerts are triggered. These alerts are indicative of risk and spot anomalies and help to take necessary steps and prevent catastrophic accidents.

In the airline industry, predictive analytics is playing vital role in airline safety events. The airline industry is working with NASA for managing the airline safety program. The NASA projects use structured and unstructured data like text analytics algorithms to extract the useful information from the huge amount of text data sets. Southwest airlines has already partnered with NASA in order to improve air travel safety.[16]

By implementing machine learning algorithms, they have built an automated system that is capable of crunching huge datasets and send early warning signals about the anomalies and prevent potential accidents. NASA is helping other commercial airlines also in improving aircraft maintenance procedures and heading off equipment failure. NASA has partnered with Honeywell Aerospace to find methods for automatically discovering precursors to contrary actions while aircraft is in flight, like predicting engine failure. NASA has also partnered with Radiometrics Corporation to improve weather forecasting and flight safety.[17] NASA Aviation Safety Investment Strategy Team (ASIST) identified in-flight icing as one of the top priorities in improving flight safety.[18,19] The crash of American Eagle Flight 4184 is one example of an in-flight icing encounter that occurred on October 31, 1994, in Roselawn, Indiana.[20,21]

Airline Fraud Detection

Fraud is rapidly increasing in the airline industry due to the nature of the transactions involved. The value of using predictive analytics and big data is increasing and helping the airline industry to detect and prevent airline fraud. Today's fast-evolving predictive technologies and machine learning solutions are able to control and prevent fraud by identifying the fraud patterns like type of fraud, origin of fraud, scope, and scale of fraud. Analysts have to stay a step ahead of fraudsters. Inadequate fraud detection mechanisms result in false alerts or failure to alert. These inadequacies reduce the robustness and reliability of the models and can make it challenging for fraud investigators to identify the occurrence of fraud. Models for fraud detection are constantly updated to incorporate new variables as well as more efficient algorithms to keep the accuracy high and identify new sources of fraud.

According IATA, airline card sales are exposed to fraud and annual loss is close to 1 billion dollars.[22] The use of online payments for purchasing airline tickets has grown enormously in the last couple of years. While doing online transactions, large amounts of personal and financial information, particularly credit data information, is stored on the Internet. Cybercriminals attack the databases, fetch the credit card information of the travelers, and sell that information to the fraudsters. These stolen credit cards are used by the fraudsters to purchase the airline tickets and cause huge financial loss to the airline industry. The industry fraud prevention project has been planned to combat airline fraud risk in the areas of card payments and cyber fraud. A recent Europol press release stated that 195 passengers were detained for airline ticket fraud.[23]

Predicting Flight Delays

Airplane travel is challenging with inconsistent flight arrival and departure times. Flight delays are not only costly for the airlines but also increase the frustration among the travelers. The airlines report the reasons of flight delays in five broad categories.[24]

1. Extreme Weather: Due to the extreme weather conditions or when weather conditions are not ideal like hurricanes, tornados, and snow storms, this makes all the flights have to change their schedule and results in flight delays and cancellations.

2. Late-Arriving Aircraft: This is the most frequent problem observed in the airline industry when the previous flight with the same aircraft arrives late, it will result in the present flight being late for departure and hence delayed flights.

3. Air Carrier: Due to aircraft maintenance problems, baggage loading, fueling, aircraft cleaning, and many more, these are the factors responsible for aircraft delays and cancellations.

4. Security: Flights delays can be caused because of security reasons like inoperative screening equipment, long queue in screening areas, reboarding of aircraft, etc.

5. National Aviation System (NAS): Due to airport operations, heavy volume of traffic, air traffic control, etc., these are the factors responsible for flight delays and cancellations in respect to NAS.

Cancelled and Diverted flights also contributed to some percentage of flights delays.

Figure 6-3 displays the statistics of overall causes of flights delays according to the Air Travel Consumer Report November 2017.[25,26]

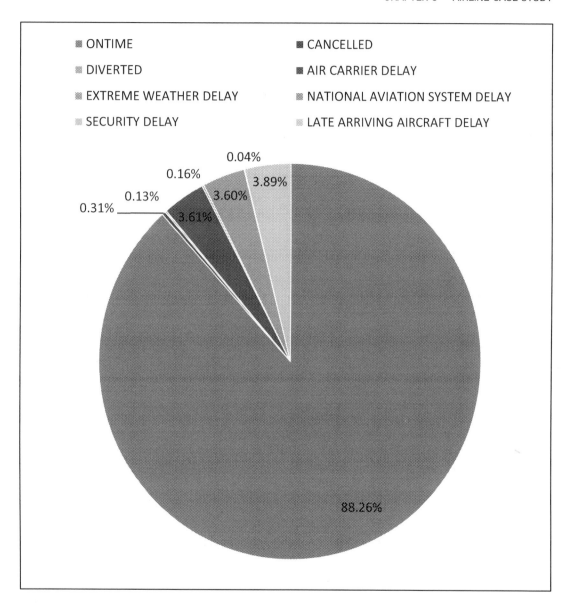

Figure 6-3. *Overall causes of flights delays*

A prediction model like the decision tree, Bayesian, Artificial Neural network, Regression, etc., are used to predict the probability of flight delay. By leveraging the power of big data, huge chunks of data can be collected from different sources easily such as flight maintenance history and flight route information, weather conditions, and security delays. Analyzing this data helps in finding the trends. Flight Deck Interval

Management (FIM) is part of NASA's Air Traffic Management Technology Demonstration -1 and their goal is to develop and evaluate new technologies and procedures related to aircraft scheduling and airport arrivals. FIM lets controllers deliver the aircraft more precisely and more predictably, which helps the airlines and airport operators to manage the air traffic more efficiently and minimize flight delays.[27] Prediction of the metrics like weather prediction, airport congestion, airport runway configuration, flight mechanical failures, etc., in advance provides actionable insights for airline industry. Early warning signals can help management to plan and adjust the flight schedule accordingly, hence mitigating flight delays and optimizing the cost.

Case Study: Predicting Flight Delays with Multiple Linear Regression Model

The rapid development of the economy results in an increase in the demand for air transport. Today's airline industry is facing lots of challenges to stay competitive in the market. Flight delay is one of the biggest challenges for all of the airlines. Due to flight delays, not only has this had a negative impact on passenger experience but also the operations of the airport are also impacted. Repeated flight delays can result in economic losses to the airlines and impact the image. Predictive analytics helps in predicting flight delays and identifying the major factors that are responsible for the repeated flight delay. An early warning signal helps the airline industry to manage their plans and flight schedules accordingly. Reduction in flight delays results in decreasing airline operation costs and provides quality and timely services to the passengers. In this case study, application of the multiple linear regression model to predict flight Arrival Delay and the major factors responsible for flight delay is demonstrated.

The Linear Regression model is an approach to estimate the linear relationship between dependent variable (Y) and one or more independent variables denoted by (x). When there is only one independent variable, then it is considered as a simple linear regression model; and in other cases when there is more than one independent variables, then it is considered a multiple linear regression model.[28] In a linear regression model, the dependent variable is continuous and independent variables can be binary, categorical, or continuous. A multiple linear regression model is used to predict the values of a dependent variable (Y) with the given value of independent variables(x_1, x_2, x_3, x_4........x_n). For example, predicting the price of the airline ticket, predicting the flight delay, predicting the airport store sales, etc., can be performed using multiple linear regression models.

Multiple Linear Regression Equation

A multiple linear regression model is used to predict the dependent variable (Y) with the given value of independent variables(x_1, x_2, x_3, x_4........x_n).[29] The relationship between the dependent and independent variables is represented by the following line equation.

$$y = \beta_0 + \beta_1 x_1 + \beta_2 x_2 + \beta_3 x_3 + \cdots \beta_n x_n + \varepsilon$$

Where y = dependent variable

β_0 = Linear regression model intercept

$x_1, x_2 \cdots x_n$ = Independent variables

$\beta_1, \beta_2 \cdots \beta_n$ = Linear regression coefficients for N independent (x) variables

ε = error term in the model

Multiple Linear Regression Assumptions and Checking for Violation of Model Assumptions

All statistical models are developed, keeping in mind certain assumptions about the characteristics of data. It is critical that data on which the model is being applied is aligned with these assumptions for obtaining reliable results. In case these assumptions are not met, the results may not be reliable. In addition to these assumptions, there are also certain model-specific formats that need to be followed to ensure the data is aligned with the fundamental workings of the model. Some of these changes can be applied when the conditioning and structuring of the data is being performed prior to model implementation.

Key multiple linear regression assumptions and checking for violation of model assumptions are listed below.[30]

1. There is a linear relationship between the dependent and independent variables. If the dependent and independent variables do not have a linear relationship, the results of regression analysis may not be reliable. A preferable method of examining this is by plotting a scatterplot of the relationship between independent variables and dependent variables. It must be checked that each scatterplot is displaying a linear relationship between the dependent and independent variables. In case the scatterplot displays the curvilinear components (squared and cubic terms), it shows that there is a nonlinear relationship between the dependent and independent variables; hence the assumptions are violated. The solution to this issue is addition of quadratic and cubic terms.

2. In linear regression, it is assumed that the residuals are normally distributed. In case the residuals are not normally distributed (highly skewed or variables with substantial outliers), it can mislead the relationships and significance tests results. In order to get the reliable results, the assumption of normal distribution must be met. This assumption can be tested by using the visual examination of data plots, skew, kurtosis, and P-P plots; also the Kolmogorov-Smirnov test is used in providing inferential statistics on normality. Outliers in that data can be identified by histograms or frequency distributions.

3. There must be homoscedasticity, meaning that the variance of the error term must be the same across all levels of independent variables. In other cases, if the variance of the error term is different at different values of independent variables, then there is the presence of heteroscedasticity leading to the poor and misleading results and the assumption is violated. This assumption can be tested by visual inspection of a standardized residual plot by the standardized regression predicted value. Ideally, when the residuals are evenly scattered around the horizontal line, there is presence of homoscedasticity; and when

the residuals are not evenly scattered around the horizontal line and takes a various shape like a bowtie, funnel shape, etc., then there is the presence of heteroscedasticity. One of the solutions to the heteroscedasticity issue is nonlinear data transformation (log transformation of variable) or addition of a quadratic term.

4. There must not be multicollinearity present between the independent variables. It means that the independent variables are not highly correlated with each other. In case there is multicollinearity present between the independent variables, then it will lead to the misleading results. This assumption can be easily tested by using a simple approach of correlation matrix for all the independent variables. If the correlations are above 0.80, then there might be the presence of multicollinearity between the independent variables. Another precise approach is variance inflation factor (VIF). If VIF is higher than 5, then there is a presence of multicollinearity and the simplest solution is to identify those variables causing multicollinearity issues through VIF values and dropping those variables from the regression.

5. The error terms or the residuals should not be correlated to each other if they are independent from each other; hence assumption is met. In case the residuals of the independent variables are correlated, then it leads to the misleading results. The residual correlation issue is mainly impacted by the sampling design. In case of simple random sampling from a large population, there is no problem of residuals correlations of the independent variables; but if the sampling method involves any type of clustering; then the problem of residuals correlations of the independent variables might exist. The solution to the residual correlation issue is quite similar to the heteroscedasticity issue. A simpler approach is to stick with least square coefficients but use robust standard errors.

Variables Selection in Multiple Linear Regression Model

In multiple linear regression modeling, there are various selection methods that specify how independent variables are allowed to enter inside the analysis.[31] By applying different selection methods, a variety of regression models from the same set of variables can be built. In a forward selection method, the independent variables are included in the model one at a time as long as they are making a significant contribution in explaining the dependent variable. This procedure is iterated until there are no independent variables left that are making any significant contribution in predicting the dependent variable. In the backward selection method, all the variables are included in the model in a single step and then each variable is removed at a time based on its least contribution in predicting the dependent variable. This procedure is iterated until not a single independent variable is left that is not making any significant contribution in predicting the dependent variable. Stepwise selection method is the combination of forward and backward selection. Like in forward selection, it begins with adding the most significant independent variable at a time that is showing the highest contribution in predicting the dependent variable and then iterates the procedure. Additionally after each step a check is done to identify whether one of the independent variables has now become irrelevant due to the relationship with other variables. If so, that particular variable is eliminated.

Evaluating the Multiple Linear Regression Model

Model fit tests are done to evaluate how well a model fits the data or how accurately the model can predict the observed values or real values. The lesser is the difference between observed values and predicted values, the better is the model. Most commonly used methods for evaluating model fit are residual plots, goodness-of-fit, and standard error of the regression model.

Residual Plots

Residual is considered as the unpredictable random component for each observation and is defined as the difference between the observed value and the predicted value. Residual from the model is one of the most informative methods to know about the model fit. The simplest way to check the randomness of the residual is by visual inspection of histograms or scatter plots of the residuals against the independent variables. In linear regression the residuals are always randomly scattered without showing any patterns. In case residuals are not randomly scattered and display some pattern, then it indicates that there is a nonlinear relationship and the linear regression model is not appropriate for such data.

Outliers Detection and Influential Observations

Extreme observations in the data, which are different from other observations in the data are considered as outliers. Observations that impact the estimation results of a linear regression model to a great extent are considered as influential observations. Outliers are always not bad for the model; sometime outliers are present in the data because of the manual error or incorrect data entry, and in such scenario those values can be removed from the model; but in some cases the outliers in the data give us useful insights and must be analyzed carefully, for example, in fraud, detection outliers are very effective in providing effective insights. The simplest way of detecting outliers is by visual inspection of residual plots, box plots, and histograms. There are some other numerical diagnostics tests or measures for detection of the outliers like Studentized residuals, Cook's distance, DFITS, and DFBETAS.[32,33]

Goodness-of-Fit

Coefficient of multiple determination or R-squared is a goodness-of-fit measure for multiple linear regression models. R-squared is the statistical measure and it indicates the percentage of the variance in the dependent variable, which is explained by independent variables.[34] Coefficient of multiple determination or R-squared summarizes how well a multiple linear regression model fits the data. Coefficient of determination or R-squared is calculated by the sum of squares due to error over the total sum of squares; or in other words, it is the variance explained by the model over total variance. The formula for calculating R-squared is mentioned below

$$R^2 = 1 - \frac{SSE}{SST}$$

Where SSE = Sum of squares due to error

SST = Total sum of squares

R-squared value is always between 0 and 100%. Zero percent denotes that the variance in the dependent variable is not explained by independent variables. One hundred percent denotes that all the variance in the dependent variable is perfectly explained by independent variables and displays a perfect linear relationship. Higher R-squared values display that there is lesser difference between the observed data and the fitted values. However, a high R-squared value does not always mean a good model and a low R-squared value does not mean a poor model for prediction. There are not any such rules that indicate what is considered as the good R-squared value, and the typical R-squared value depends upon the type of the data used for regression analysis. The issue with the R-squared value is that it values increases with the addition of independent variables and can mislead the results. In order to overcome this issue, adjusted R-squared statistics are used. The adjusted R-squared adjusts for the number of parameters in the model, and its value increases when the addition of new parameter shows some improvement in the model fit; or else it remains same. Hence, an adjusted R-squared statistic is more preferred than a R-squared statistic.

Standard Error of the Regression Model

Standard error of the regression is also known as standard error of the estimate and is another numeric measure that tells that how well the model fits the sample data. In standard error of the regression, the absolute measure of the distance that the data points fall from the regression line is calculated. This statistic tells us how far are the data points from the regression line on average. Standard error of regression is in units of dependent variable. The lower is the value of the standard error, the more precise is the model prediction. A lower value of standard error of regression signifies that there is a small distance between the data points and fitted values.

Multiple Linear Regression Model Using R

In this Airline case study:

   Business Problem: To predict the flight arrival delays

   Business Solution: To build the multiple linear regression model

About Data

In this airline case study flight_delay data set is generated synthetically and is used to develop the multiple linear regression model in order to predict flight Arrival Delays. The flight_delay data set contains total 3,593 observations and 11 variables; 10 variables are numeric and 1 variable is categorical. Arr_Delay is the dependent or target variable that is continuous in nature and all other variables are independent or explanatory variables in the data.

```
#Read the data from the working directory, create your own working directly
to read the dataset.

setwd("C:/Users/Deep/Desktop/data")

data1 <- read.csv ("C:/Users/Deep/Desktop/data/
flight_delay_data.csv",header=TRUE,sep=",")
```

Performing Data Exploration

```
#perform exploratory data analysis to know about the data

# display top 6 rows of dataset to see how data look like

head (data1)
```

	Carrier	Airport_Distance	Number_of_flights	Weather	Support_Crew_Available
1	UA	437	41300	5	83
2	UA	451	41516	5	82
3	AA	425	37404	5	175
4	B6	454	44798	6	49
5	DL	455	40643	6	55
6	UA	416	39707	5	146

	Baggage_loading_time	Late_Arrival_o	Cleaning_o	Fueling_o	Security_o	Arr_Delay
1	17	19	15	26	31	58
2	17	19	15	22	32	48
3	16	17	14	28	29	16
4	18	19	13	29	31	81
5	17	19	18	26	37	62
6	16	19	6	28	31	34

```
# display bottom 6 rows

tail(data1)
```

	Carrier	Airport_Distance	Number_of_flights	Weather	Support_Crew_Available
3588	B6	431	43641	5	79
3589	UA	442	41793	6	120
3590	EV	453	42219	5	70
3591	DL	427	41847	6	49
3592	B6	395	38732	5	74
3593	B6	450	46906	5	51

	Baggage_loading_time	Late_Arrival_o	Cleaning_o	Fueling_o	Security_o	Arr_Delay
3588	16	18	10	22	37	35
3589	17	19	8	26	42	57
3590	16	19	12	20	37	43
3591	16	19	4	31	42	45
3592	17	18	10	23	39	40
3593	18	20	9	27	37	92

```
# describe the structure of data

str(data1)
'data.frame'          : 3593 obs. of  11 variables:
 $ Carrier              : Factor w/ 14 levels "9E","AA","AS",..:
                          11 11 2 4 5 11 4 6 4 2 ...
 $ Airport_Distance     : int   437 451 425 454 455 416 439 446 441 456 ...
```

```
$ Number_of_flights      : int   41300 41516 37404 44798 40643 39707 45627
                                 40415 42248 43453 ...
$ Weather                : int   5 5 5 6 6 5 5 5 6 6 ...
$ Support_Crew_Available: int   83 82 175 49 55 146 141 145 58 3 ...
$ Baggage_loading_time   : int   17 17 16 18 17 16 16 17 17 18 ...
$ Late_Arrival_o         : int   19 19 17 19 19 19 19 18 19 19 ...
$ Cleaning_o             : int   15 15 14 13 18 6 15 9 9 10 ...
$ Fueling_o              : int   26 22 28 29 26 28 21 25 17 24 ...
$ Security_o             : int   31 32 29 31 37 31 39 39 53 36 ...
$ Arr_Delay              : int   58 48 16 81 62 34 63 47 39 122 ...
```

#display the column name of the data

```
names(data1)
 [1] "Carrier"            "Airport_Distance"      "Number_of_flights"
 [4] "Weather"            "Support_Crew_Available" "Baggage_loading_time"
 [7] "Late_Arrival_o"     "Cleaning_o"            "Fueling_o"
[10] "Security_o"         "Arr_Delay"
```

#display the summary or descriptive statistics of the data

```
summary(data1$Arr_Delay)

   Min.  1st Qu.  Median   Mean   3rd Qu.    Max.
   0.0    49.0     70.0    69.8    90.0     180.0
```

#Let's check the missing values present in the data

```
sum(is.na(data1))
[1] 0
```

#To find out the correlation between the variables

```
corr <- cor.test(data1$Arr_Delay,data1$Number_of_flights, method = "pearson" )
corr
```

```
Pearson's product-moment correlation
data:   data1$Arr_Delay and data1$Number_of_flights
t = 86.823, df = 3591, p-value < 2.2e-16
alternative hypothesis: true correlation is not equal to 0
95 percent confidence interval:
 0.8121611 0.8332786
sample estimates:
     cor
0.823004
```

Pearson moment correlation method is used to find out the correlation between the two variables. There is a strong positive correlation (82%) between Arr_Delay and Number_of_Flights, which means both of the variables are directly proportional to each other. Arr_Delay increases with the increase in Number_of_Flights. To determine whether the correlation between variables is significant, we need to compare the p-value to the significance level (0.05). In this case the p-value for the correlation between Arr_Delay and Number_of_Flights is less than the significance level of 0.05, which indicates that the correlation coefficient is significant.

```
#To add four charts in one window or plotting panel par(mfrow = c(2,2))
#To plot the dependent and independent variable

plot(data1$Arr_Delay,data1$Number_of_flights)

plot(data1$Arr_Delay,data1$Security_o)

plot(data1$Arr_Delay,data1$Support_Crew_Available)

plot(data1$Arr_Delay,data1$Airport_Distance)
```

Figure 6-4 displays the various plots of dependent and independent variable.

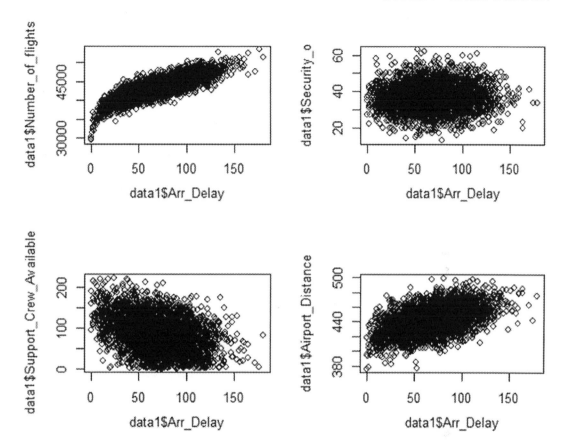

Figure 6-4. *Plotting the Variable*

```
#To drop the first variable (carrier) from data1

data2 <- data1[-c(1)]
```

Carrier variable is removed from data1 as it is categorical in nature and not fit for applying cor function.

```
#To check the correlation between the variables

cor(data2)
```

	Airport_Distance	Number_of_flights	Weather	Support_Crew_Available
Airport_Distance	1.000000000	0.40561880	0.148396273	-0.17006177
Number_of_flights	0.405618798	1.00000000	0.247261335	-0.30060555
Weather	0.148396273	0.24726133	1.000000000	-0.11068882
Support_Crew_Available	-0.170061773	-0.30060555	-0.110688820	1.00000000
Baggage_loading_time	0.400131114	0.67192355	0.271090293	-0.29906280
Late_Arrival_o	0.317892412	0.57064820	0.219533717	-0.24226314
Cleaning_o	-0.003433746	-0.01181124	0.006311967	-0.02756499
Fueling_o	-0.037275014	-0.04282957	0.011344078	0.03805393
Security_o	0.043956588	0.07465514	0.034967545	-0.01520496
Arr_Delay	0.482167140	0.82300401	0.327185319	-0.36188832

	Baggage_loading_time	Late_Arrival_o	Cleaning_o	Fueling_o
Airport_Distance	0.40013111	0.31789241	-0.003433746	-0.037275014
Number_of_flights	0.67192355	0.57064820	-0.011811241	-0.042829570
Weather	0.27109029	0.21953372	0.006311967	0.011344078
Support_Crew_Available	-0.29906280	-0.24226314	-0.027564994	0.038053928
Baggage_loading_time	1.00000000	0.54694337	-0.011580701	-0.004914550
Late_Arrival_o	0.54694337	1.00000000	-0.016692814	-0.020889830
Cleaning_o	-0.01158070	-0.01669281	1.000000000	-0.007044785
Fueling_o	-0.00491455	-0.02088983	-0.007044785	1.000000000
Security_o	0.07395267	0.08105316	-0.011456912	0.016647225
Arr_Delay	0.78363370	0.66683947	-0.003494777	-0.036260638

	Security_o	Arr_Delay
Airport_Distance	0.04395659	0.482167140
Number_of_flights	0.07465514	0.823004007
Weather	0.03496754	0.327185319
Support_Crew_Available	-0.01520496	-0.361888320
Baggage_loading_time	0.07395267	0.783633696
Late_Arrival_o	0.08105316	0.666839470
Cleaning_o	-0.01145691	-0.003494777
Fueling_o	0.01664723	-0.036260638
Security_o	1.00000000	0.080079308
Arr_Delay	0.08007931	1.000000000

```
#Splitting dataset into training and testing dataset
```

Data is split into two parts train_data set and test_data set by using stratified random sampling, and the splitting ratio is 70:30, it means 70% of data contribute to the train dataset and 30% of data contribute to test dataset. Train_data set is used to build the model and test_ data set is used to test the performance of the model.

```
#Install caTools package for splitting the data

install.packages("caTools")

library(caTools)

#To reproduce the sample

set.seed(1000)

sample <- sample.split(data2$Arr_Delay,SplitRatio=0.70)

#split of the data using subset command

train_data <- subset(data2,sample==TRUE)
test_data <- subset(data2,sample==FALSE)
```

Model Building & Interpretation on Training and Testing Data

Train_data set is used to build the model and test_ data set is used to test the performance of the model.

```
#Building multiple linear regression model using lm on train_data set

model <- lm(Arr_Delay ~., data = train_data)
summary(model)
```

In the above code lm function (linear model) is used to build the multiple linear regression model. Arr_Delay is the dependent or target variable in the data and all other variables are independent or explanatory variables. Train_data set is used for building the model and summary model will display the coefficients table and model statistics. Listing 6-1 displays the coefficients table and model statistics.

Looking at the coefficients in Listing 6-1, there are variables whose p-values are less than 0.001, hence all those variables are the significant variables in the model, but there are variables like Cleaning_o, Fueling_o and Security_o whose p-values are not less than 0.001; hence these variables are not significant variable in the model. Let's remove all insignificant variables from the model and run the model again.

Listing 6-1. Coeffiecients Table

```
Call:
lm(formula = Arr_Delay ~., data = train_data)

Residuals:
    Min      1Q  Median      3Q     Max
-35.780  -8.151  -0.583   8.230  70.332

Coefficients:
                            Estimate Std. Error t value Pr(>|t|)
(Intercept)               -5.801e+02  8.943e+00 -64.861  < 2e-16 ***
Airport_Distance           1.751e-01  1.613e-02  10.852  < 2e-16 ***
Number_of_flights          4.418e-03  1.281e-04  34.486  < 2e-16 ***
Weather                    4.721e+00  5.403e-01   8.737  < 2e-16 ***
Support_Crew_Available    -5.113e-02  6.398e-03  -7.991 2.03e-15 ***
Baggage_loading_time       1.353e+01  5.238e-01  25.835  < 2e-16 ***
Late_Arrival_o             6.999e+00  3.931e-01  17.804  < 2e-16 ***
Cleaning_o                 1.176e-01  7.106e-02   1.654   0.0982 .
Fueling_o                 -1.016e-01  7.086e-02  -1.433   0.1519
Security_o                 8.947e-03  3.505e-02   0.255   0.7985
---
Signif. codes:  0 '***' 0.001 '**' 0.01 '*' 0.05 '.' 0.1 ' ' 1

Residual standard error: 12.38 on 2504 degrees of freedom
Multiple R-squared:  0.8213,    Adjusted R-squared:  0.8206
F-statistic:  1279 on 9 and 2504 DF,  p-value: < 2.2e-16
```

```
#Building final multiple linear regression model with significant
variables  on train_data set

model_sig<-lm(Arr_Delay~Airport_Distance+Number_of_flights
+Weather+Support_Crew_Available+Baggage_loading_time
+Late_Arrival_o, data= train_data)

model_sig
```

Listing 6-2. Coeffiecients Table

```
Call:
lm(formula = Arr_Delay ~ Airport_Distance + Number_of_flights +
    Weather + Support_Crew_Available + Baggage_loading_time +
    Late_Arrival_o, data = train_data)

Coefficients:
          (Intercept)        Airport_Distance        Number_of_flights
            -5.809e+02                1.754e-01                4.424e-03
               Weather   Support_Crew_Available     Baggage_loading_time
            4.688e+00               -5.171e-02                1.350e+01
        Late_Arrival_o
            7.007e+00

summary(model_sig)

Call:
lm(formula = Arr_Delay ~ Airport_Distance + Number_of_flights +
    Weather + Support_Crew_Available + Baggage_loading_time +
    Late_Arrival_o, data = train_data)
```

```
Residuals:
    Min      1Q  Median      3Q      Max
-36.224  -8.213  -0.677   8.295   70.468

Coefficients:
                         Estimate Std. Error t value Pr(>|t|)
(Intercept)            -5.809e+02  8.705e+00 -66.734  < 2e-16 ***
Airport_Distance        1.754e-01  1.613e-02  10.877  < 2e-16 ***
Number_of_flights       4.424e-03  1.280e-04  34.570  < 2e-16 ***
Weather                 4.688e+00  5.400e-01   8.683  < 2e-16 ***
Support_Crew_Available -5.171e-02  6.394e-03  -8.087 9.38e-16 ***
Baggage_loading_time    1.350e+01  5.232e-01  25.804  < 2e-16 ***
Late_Arrival_o          7.007e+00  3.930e-01  17.828  < 2e-16 ***
---
Signif. codes:  0 '***' 0.001 '**' 0.01 '*' 0.05 '.' 0.1 ' ' 1

Residual standard error: 12.39 on 2507 degrees of freedom
Multiple R-squared:  0.8209,    Adjusted R-squared:  0.8205
F-statistic:  1916 on 6 and 2507 DF,  p-value: < 2.2e-16
```

Listing 6-2 shows the coefficients (labeled Estimates), their standard errors, the t values and the P-values. The coefficients for Airport_Distance, Number_of_flights, Weather, Support_Crew_Available, Baggage_loading_time and Late_Arrival_o are statistically significant as their P-values are less than 0.001. In a linear regression model, for a one-unit increase in the predictor or independent variables, it will display the change (increase or decrease) in the outcome or dependent variable. It can be explained as:

1. For every one-unit increase in Number_of_flights increases the Arr_Delay by **(0.004)**

2. Similarly it is for Support_Crew_Available that one-unit decrease in Support_Crew_Available increases the Arr_Delay by **(-0.05)**

3. Similarly it is for Late_Arrival_o that one-unit increase in Late_Arrival_o increases the Arr_Delay by **(7.007)**

In this model Residual Standard error is 12.39 (2 *12.39=24.78) on 2507 degrees of freedom Multiple R-squared is 82.09% and Adjusted R-squared is 82.05%. An R-squared value 0.7 (70%) or higher is generally accepted as good. In this case study R-squared is 82.09%, it means that 82.09% of the flight Arr_Delay is predicted by the independent variables in our model like Airport_Distance, Weather, Number_of_flights, Support_Crew_Available, Baggage_loading_time and Late_Arrival_o. There is a limitation of R-squared as it has the tendency to increase with the increase in the number of the variables in the model; therefore, Adjacent R-squared is preferred when there are more numbers of variables. Standard error of the regression and Multiple R-squared is considered as the Goodness-of-fit measures in explaining how well a linear regression model fits the sample data. F test (P-value <0.001) indicates that the model is significant for predicting Arr_Delay based on a group of independent variables in the model (Full model is better than Reduced model). Detailed explanation about Standard error of the regression, Multiple R-squared and Adjusted R-squared are provided in the above section.

```
names(model_sig)
```

```
 [1] "coefficients"  "residuals"   "effects"      "rank"      "fitted.values"
 [6] "assign"        "qr"          "df.residual"  "xlevels"   "call"
[11] "terms"         "model"
```

```
#number of the fitted values
```

```
length(model_sig$fitted.values)
[1] 2514
```

The number of fitted values is 2514 and the number of observations in train_data set is also 2514; it means that the fitted values are the predicted values of the train_data set.

```
#predicting fitted values of train_data set
pred_train<- model_sig$fitted.values
head(pred_train)
```

```
       1         3         4        10        11        12
60.25684   8.64335  98.66114  95.44036  72.93578 104.41100
```

```
pred_train1 <- data.frame(pred_train)
```

```
#residual values
resed_train <- model_sig$residuals
head(resed_train)
```

```
         1           3           4          10          11          12
 -2.256842    7.356650 -17.661140   26.559637 -13.935782   17.589005
```

```
resed_train1<-data.frame(resed_train)
```

Residual values are the difference between the actual values and the predicted values. If actual values are greater than predicted values, then the residual values are positive and if actual values are smaller than predicted values, then the residual values are negative.

```
# Prediction on the unforeseen dataset i.e on  test data
```

```
pred_test<- predict(model_sig,newdata = test_data)
head(pred_test)
```

```
      2         5        6         7        8        9
63.71968 66.64359 32.76744 63.25060 47.70671 71.13357
```

```
pred_test1<- data.frame(pred_test)
```

```
#Plotting Actual versus Predicted outcome
```

```
plot (test_data$Arr_Delay,col="red",type ="l",lty=1.8)
lines(pred_test1,col="blue",type ="l",lty=1.4)
```

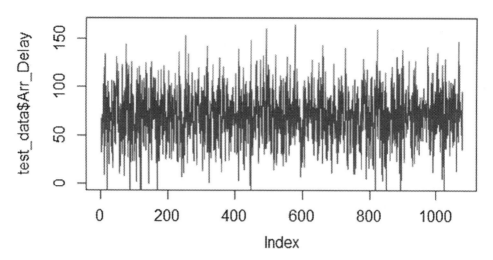

Figure 6-5. *Plot of Actual versus Predicted outcome*

Figure 6-5 displays the plot of actual versus predicted outcome. Red lines represent the actual values from test-data set and blue lines represent the predicted values. From the above graph it is seen that most of the number of cases of blue and red lines are overlapping with each other except in a few cases. In this case it is seen that most of the predicted values are closely following the actual values, indicating it the quiet, good model.

```
#Linear Regression Diagnostics statistics and plot
```

After the model-building adequacy of the model is checked by looking at regression results like Slope coefficients, P-values, R-squared, Adjacent R-squared Standard error and F test statistics, it is not enough to stop your regression analysis model. It is always the best approach to look at the linear regression assumptions that assumptions are not violated before submitting the final findings of the model. Violation of the assumptions can lead to the unreliable and inaccurate inferences and predictions. A few diagnostics statistics and plots like residual, normality, etc., are mentioned below.

```
# plot function to view four different diagnostic plots
```

Plot function is used to see four different diagnostic plots. These plots are used for visual inspection to see how the model did on linear regression assumptions outlined in the multiple linear regression assumptions.

```
# To view first plot
  plot(model_sig,which=1)
```

The first plot is Residual versus Fitted and is displayed in Figure 6-6, showing the residual versus fitted values and it is used for checking the linearity and homoscedasticity assumptions. If the linear model assumption is not satisfied, then the residuals display some shape or pattern like a parabola and it is bad. The scatterplot of residual should not display any pattern and the residuals must be equally spread around the y = 0 red line. It is seen from the below plot that three data points that have large residuals are flagged (observations 1647, 1652, 3163). Besides these, our residuals appear to be ok and evenly spread around y = 0 red line.

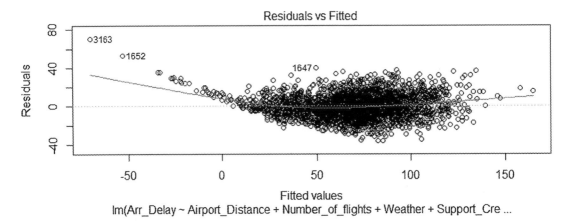

Figure 6-6. *Plot of Residual versus Fitted values*

```
# Let's look at our second plot now.

plot(model_sig,which=2)
```

The second plot is QQ-plot and is displayed in Figure 6-7. QQ-plot is used for checking the normality assumptions that the residuals are normally distributed if the residual deviates from a greater extent from the straight dashed line. then it is considered bad; and if the residuals appear to be lined well on the straight dashed line, it is good. It is seen from the below plot that same three data points that have large residuals are flagged (observations 1647, 1652, 3163). Besides these three data points, residuals appear to be lined well on the straight dashed line.

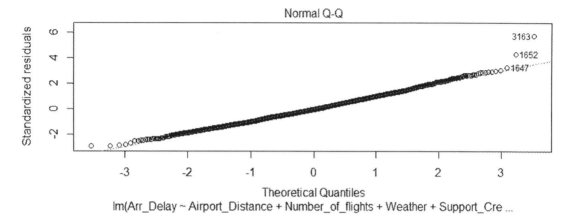

Figure 6-7. *QQ-plot*

```
# Let's look at our third plot now.

plot(model_sig,which=3)
```

The third plot is a Scale Location plot and is displayed in Figure 6-8 where it shows the square rooted standardized residuals versus fitted values and is used for checking the homoscedasticity assumptions. If the residuals are not randomly spread and the red line is not horizontal, then it is not good; and if the red line is horizontal with randomly spread data points, then it is good. From the below plot it is seen that the same three data points that have large residuals are flagged (observations 1647, 1652, 3163). Besides these, residuals appear randomly spread around the red horizontal line assuming that homoscedasticity assumptions hold here.

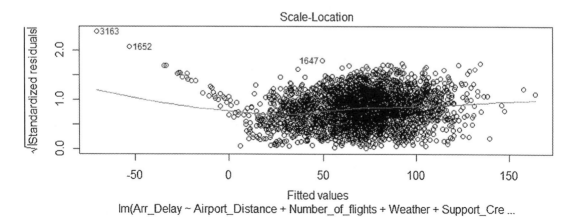

Figure 6-8. *Scale Location plot*

```
# Let's look at our final plot(fourth) now.
#Note that this specific plot ID is considered '5' in R.hence    mentioned
as which=5.
```

```
 plot(model_sig,which=5)
```

The fourth plot is displayed in Figure 6-9 and is used for finding the influential cases, if any such cases are present in the data. There are extreme values or outliers present in the data, but not all outliers are influential in describing a linear regression line. It means that regression results would not be much different if those outliers are either excluded or included from regression analysis and follows the trend as they are not influential. On the other side, if there are some outliers that are influential cases, then they could impact a linear regression line and can change the regression results severely, once excluded from the analysis.

If there are no dash red curve lines or Cook's distance curve lines, it is good and there are no influential cases. In case there are dash red curve lines and none of the data points are within Cook's distance, then it means those cases are not influential; but if any of the data points cross the Cook's distance line; then it is not good and there are influential cases present in the data. From the below plot it is seen that that three data points that have large residuals are flagged (observations 1539, 1652, 3163). However, none of these data points cross Cook's distance line and are not even present in this plot, so there are no influential cases.

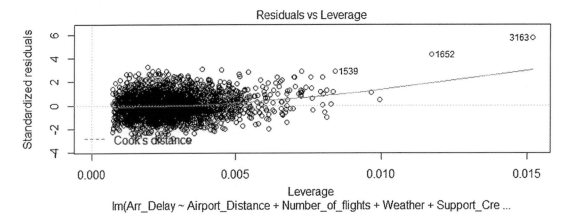

Figure 6-9. *Influential plot*

From all the above four diagnostic plots, observations 1647, 1652, and 3163 are flagged. All these three observations must be analyzed closely in order to find out the reasons that can be due to data entry, data collection, etc. If it is because of data entry, it must be excluded. In this case all three data points (1647, 1652, 3163) are removed from the data and when the model is run again, there are no severe changes in the regression results. Besides these three data points (1647, 1652, 3163), our data look good.

```
#Install car package

install.packages("car")

library(car)

# Test for Independence Assumption

durbinWatsonTest(model_sig)
```

The Durbin-Watson test is a measure of autocorrelation in residuals from regression analysis. If the residuals are autocorrelated, the regression results can be inaccurate and reliable. The Durbin-Watson test value ranges from 0 to 4.

1. 2 means no autocorrelation

2. >2 to 4 is negative autocorrelation

3. 0 to <2 is positive autocorrelation

In the Durbin-Watson test the null hypothesis (H0) is that there is no first order autocorrelation and an alternate hypothesis (H1) is that there is existence of first order correlation (lag is one-time unit for a first order).

```
 lag Autocorrelation D-W Statistic p-value
  1      -0.01063148     2.020454   0.642
 Alternative hypothesis: rho != 0
```

In this case $P > 0.001$ (0.642), hence accepting the null hypothesis, so there is no autocorrelation in residuals and the independence assumption is satisfied.

```
# Statistic Test for Homoscedasticity  Assumption

 ncvTest(model_sig)
```

The ncvTest is the statistic test for homoscedasticity assumption. This test helps in identifying whether there is any non-constant variance in the residual (heteroscedasticity). In the ncvTest the null hypothesis (H0) is that there is constant variance and alternate hypothesis (H1) is that there is no constant variance.

```
Non-constant Variance Score Test
Variance formula: ~ fitted.values
Chisquare = 0.9386854    Df = 1    p = 0.3326162
```

In this case P > 0.001 (0.3326162), hence accepting the null hypothesis, so there is constant variance and the homoscedasticity assumption is satisfied.

```
#VIF Test for Collinearity

vif(model_sig)
        Airport_Distance      Number_of_flights
               1.234012               2.091616
                Weather    Support_Crew_Available
               1.100304               1.128461
     Baggage_loading_time          Late_Arrival_o
               2.070252               1.600201

sqrt(vif(model_sig)) > 5

        Airport_Distance      Number_of_flights
                   FALSE                  FALSE
                Weather    Support_Crew_Available
                   FALSE                  FALSE
     Baggage_loading_time          Late_Arrival_o
                   FALSE                  FALSE
```

Variance Inflation factor (VIF) test is a measure of multicollinearity. VIF values range between 5 to 10. If in a model the VIF value of variables is greater than 5, then there is multicollinearity; and if the VIF value is greater than 10, then there is severe multicollinearity issues. In order to fix this issue, there are various solutions like highly correlated independent variables must be dropped from the model, independent variables can be standardized, adding the independent variables, etc. Severe multicollinearity does not impact the model fit but it creates lot of other problems, like it saps the statistical power of analysis, as it can increase the variance of the coefficients estimates, can change the signs of coefficients and makes it very difficult to choose the correct model. In this case, none of the independent variables are having VIF>5, hence there is no multicollinearity present in the model.

Multiple Linear Regression Model Using SAS

In this section, we discuss about different SAS procedures like proc content, proc means, proc corr, and proc univariate. We also discuss about building multiple linear regression models for predicting flight arrival delays in minutes with an explanation of program1 SAS codes and output of each part.

```
/*Create your own library in SAS like here it is libref and mention the
path*/
libname libref "/home/aro1260/deep";
/*Importing flight_delay dataset */
PROC IMPORT DATAFILE= "/home/aro1260/data/flight_delay.csv"
     DBMS=CSV Replace
     OUT=libref.flight_delay;
     GETNAMES=YES;
RUN;

/*To check the contents of the data*/
PROC CONTENTS DATA=libref.flight_delay;
RUN;
```

Partial output of proc content is displayed in Table 6-1 of part 1 and part 2 based on their importance in terms of analysis. Proc content displays the information about the data like the number of observations, number of variables, library name, and data type of each variable with their length, format, and informat.

Table 6-1.

Part 1

The CONTENTS Procedure			
Data Set Name	LIBREF.FLIGHT_DELAY	**Observations**	3593
Member Type	DATA	**Variables**	11
Engine	V9	**Indexes**	0
Created	05/15/2018 12:58:24	**Observation Length**	88
Last Modified	05/15/2018 12:58:24	**Deleted Observations**	0
Protection		**Compressed**	NO
Data Set Type		**Sorted**	NO
Label			
Data Representation	SOLARIS_X86_64, LINUX_X86_64, ALPHA_TRU64, LINUX_IA64		
Encoding	utf-8 Unicode (UTF-8)		

Part 2

Alphabetic List of Variables and Attributes					
#	**Variable**	**Type**	**Len**	**Format**	**Informat**
2	Airport_Distance	Num	8	BEST12.	BEST32.
11	Arr_Delay	Num	8	BEST12.	BEST32.
6	Baggage_loading_time	Num	8	BEST12.	BEST32.
1	Carrier	Char	2	$2.	$2.
8	Cleaning_o	Num	8	BEST12.	BEST32.
9	Fueling_o	Num	8	BEST12.	BEST32.
7	Late_Arrival_o	Num	8	BEST12.	BEST32.
3	Number_of_flights	Num	8	BEST12.	BEST32.
10	Security_o	Num	8	BEST12.	BEST32.
5	Support_Crew_Available	Num	8	BEST12.	BEST32.
4	Weather	Num	8	BEST12.	BEST32.

```
/*Descriptive statistics of the data*/

proc means data = libref.flight_delay;
vars Airport_Distance Number_of_flights Weather
Baggage_loading_time Late_Arrival_o;
run;
```

Applying proc means that the descriptive statistic or summary of the data is displayed like Number of observations (N), Mean, Standard deviation, and Min and Max values of the respective variable are represented.

Table 6-2.

The MEANS Procedure

Variable	N	Mean	Std Dev	Minimum	Maximum
Airport_Distance	3593	442.3568049	17.1662299	376.0000000	499.0000000
Number_of_flights	3593	43310.91	2797.43	29475.00	53461.00
Weather	3593	5.3534651	0.4781121	5.0000000	6.0000000
Baggage_loading_time	3593	16.9788478	0.6827480	14.0000000	19.0000000
Late_Arrival_o	3593	18.7400501	0.7934179	15.0000000	22.0000000

```
/*To check the Correlation of the dependent and independent variables in
the data*/

proc corr data = libref.flight_delay nosimple;
var Arr_Delay Airport_Distance Number_of_flights Weather Support_Crew_Available
Baggage_loading_time Late_Arrival_o Cleaning_o     Fueling_o Security_o;
run;
```

Proc corr displays the correlation matrix between dependent and independent variables. It is the best practice to look at the correlation value before model building.

Table 6-3.

The CORR Procedure

10 Variables: Arr_Delay Airport_Distance Number_of_flights Weather Support_Crew_Available Baggage_loading_time Late_Arrival_o Cleaning_o Fueling_o Security_o

Pearson Correlation Coefficients, N = 3593 Prob > |r| under H0: Rho=0

	Arr_Delay	Airport_Distance	Number_of_flights	Weather	Support_Crew_Available	Baggage_loading_time	Late_Arrival_o	Cleaning_o	Fueling_o	Security_o
Arr_Delay	1.00000	0.48217 <.0001	0.82300 <.0001	0.32719 <.0001	-0.36189 <.0001	0.78363 <.0001	0.66684 <.0001	-0.00349 0.8341	-0.03626 0.0297	0.08008 <.0001
Airport_Distance	0.48217 <.0001	1.00000	0.40562 <.0001	0.14840 <.0001	-0.17006 <.0001	0.40013 <.0001	0.31789 <.0001	-0.00343 0.8370	-0.03728 0.0255	0.04396 0.0084
Number_of_flights	0.82300 <.0001	0.40562 <.0001	1.00000	0.24726 <.0001	-0.30061 <.0001	0.67192 <.0001	0.57065 <.0001	-0.01181 0.4791	-0.04283 0.0102	0.07466 <.0001
Weather	0.32719 <.0001	0.14840 <.0001	0.24726 <.0001	1.00000	-0.11069 <.0001	0.27109 <.0001	0.21953 <.0001	0.00631 0.7053	0.01134 0.4967	xplian0.03497 0.0361
Support_Crew_Available	-0.36189 <.0001	-0.17006 <.0001	-0.30061 <.0001	-0.11069 <.0001	1.00000	-0.29906 <.0001	-0.24226 <.0001	-0.02756 0.0985	0.03805 0.0225	-0.01520 0.3622
Baggage_loading_time	0.78363 <.0001	0.40013 <.0001	0.67192 <.0001	0.27109 <.0001	-0.29906 <.0001	1.00000	0.54694 <.0001	-0.01158 0.4877	-0.00491 0.7684	0.07395 <.0001
Late_Arrival_o	0.66684 <.0001	0.31789 <.0001	0.57065 <.0001	0.21953 <.0001	-0.24226 <.0001	0.54694 <.0001	1.00000	-0.01669 0.3172	-0.02089 0.2106	0.08105 <.0001
Cleaning_o	-0.00349 0.8341	-0.00343 0.8370	-0.01181 0.4791	0.00631 0.7053	-0.02756 0.0985	-0.01158 0.4877	-0.01669 0.3172	1.00000	-0.00704 0.6729	-0.01146 0.4924
Fueling_o	-0.03626 0.0297	-0.03728 0.0255	-0.04283 0.0102	0.01134 0.4967	0.03805 0.0225	-0.00491 0.7684	-0.02089 0.2106	-0.00704 0.6729	1.00000	0.01665 0.3185
Security_o	0.08008 <.0001	0.04396 0.0084	0.07466 <.0001	0.03497 0.0361	-0.01520 0.3622	0.07395 <.0001	0.08105 <.0001	-0.01146 0.4924	0.01665 0.3185	1.00000

```
/*Applying proc univariate for detailed summary statistics*/

proc univariate data= libref.flight_delay plot;
var Late_Arrival_o;
run;
```

Proc univariate is one of the procedures that is used to display the detailed summary or descriptive statistic of the data. It will display kurtosis, skewness, standard deviation, uncorrected ss, corrected ss, standard error mean, variance, range, interquartile range, etc. It is also used to assess the normal distribution of the data and help in detecting the outliers or extreme values present in the data. Figure 6-10 displays the plots (Histogram, Boxplot, and QQ-plot) for the Late_Arrival_o variable present in the data. Extreme values for the Late_Arrival_o variable are clearly visible in all the below plots, and in QQ-plot there is no deviation seen near the tails; hence normality exists.

Part 1

The UNIVARIATE Procedure

Variable: Late_Arrival_o

Moments

N	3593	Sum Weights	3593
Mean	18.7400501	Sum Observations	67333
Std Deviation	0.79341787	Variance	0.62951191
Skewness	-0.1915608	Kurtosis	0.28347494
Uncorrected SS	1264085	Corrected SS	2261.20679
Coeff Variation	4.23380868	Std Error Mean	0.01323651

Part 2

Basic Statistical Measures

Location		Variability	
Mean	18.74005	Std Deviation	0.79342
Median	19.00000	Variance	0.62951
Mode	19.00000	Range	7.00000
		Interquartile Range	1.00000

Part 3

Tests for Location: Mu0=0				
Test		Statistic	p Value	
Student's t	T	1415.785	Pr > ltl	<.0001
Sign	M	1796.5	Pr >= IMI	<.0001
Signed Rank	S	3228311	Pr >= ISI	<.0001

Part 4

Quantiles (Definition 5)	
Level	Quantile
100% Max	22
99%	20
95%	20
90%	20
75% Q3	19
50% Median	19
25% Q1	18
10%	18
5%	17
1%	17
0% Min	15

Part 5

Extreme Observations			
Lowest		Highest	
Value	Obs	Value	Obs
15	3163	21	2037
16	3459	21	2238
16	3439	21	2951
16	3389	21	3127
16	3121	22	1047

Part 6

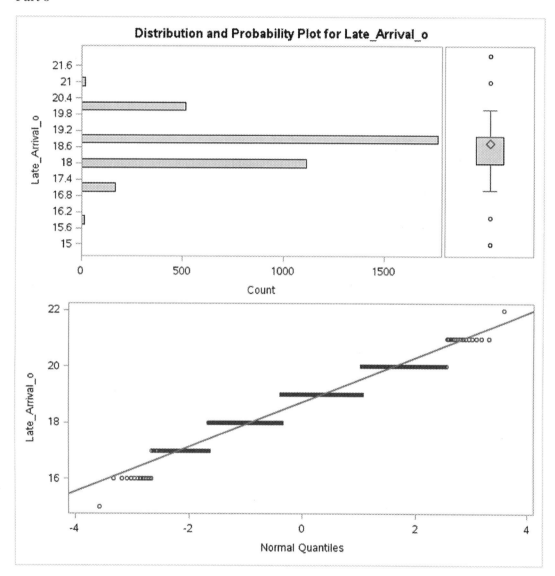

Figure 6-10. *Histogram, Boxplot, and QQ-plot displaying extreme values*

```
/*To check the relationship between the variables*/

proc sgscatter data = libref.flight_delay;
plot Arr_Delay*(Airport_Distance Number_of_flights Weather Support_Crew_
Available Baggage_loading_time Late_Arrival_o Cleaning_o Fueling_o Security_o);
run;
```

The SGSCATTER procedure is used to create a paneled graph of scatter plots for multiple variables. Figure 6-11 displays the paneled graph of scatter plots, showing the relationship between the dependent and independent variables in the flight_delay data. For example, there is a positive relationship between Arr_Delay and Airport_Distance and a negative relationship between Arr_Delay and Support_Crew_Available.

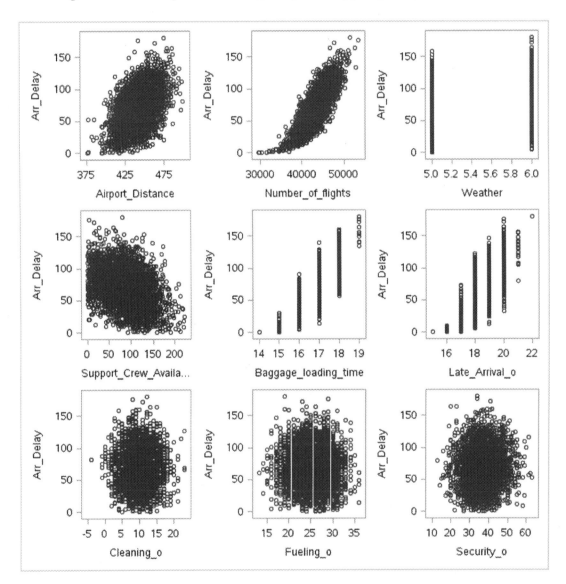

Figure 6-11. *Paneled graph of scatter plots for multiple variables*

```
/*sorting the data set by strata variable (Arr_Delay)*/

proc sort data=libref.flight_delay;
by Arr_Delay;
run;
```

In the Survey select Procedure section, we split data by a random sampling method and sampling rate as 0.7 and seed is set at 1000 (same seed value used in R). We are setting seed value to reproduce the same sample, and libref.flight_delay_subset is the output dataset after sampling is conducted.

```
/*splitting datasets into training (70%) and testing (30%)*/
proc surveyselect data= libref.flight_delay  method=srs seed=1000 outall
samprate=0.7 out= libref.flight_delay_subset;
strata Arr_Delay ;
run;
```

Table 6-4.

The SURVEYSELECT Procedure	
Selection Method	Simple Random Sampling
Strata Variable	Arr_Delay
Input Data Set	FLIGHT_DELAY
Random Number Seed	1000
Stratum Sampling Rate	0.7
Number of Strata	159
Total Sample Size	2597
Output Data Set	FLIGHT_DELAY_SUBSET

```
/*Printing first 20 observations of libref.flight_delay_subset output*/

proc print data=libref.flight_delay_subset(obs=20);
run;
```

Table 6-5.

Obs	Selected	Arr_Delay	Carrier	Airport_Distance	Number_of_flights	Weather	Support_Crew_Available	Baggage_loading_time	Late_Arrival_o	Cleaning_o	Fueling_o	Security_o	SelectionProb	Sampling Weight
1	1	0	MQ	395	29475	5	190	15	16	10	24	35	1.00000	1.00
2	1	0	FL	418	30300	5	158	14	16	5	26	27	1.00000	1.00
3	1	0	B6	376	29777	5	93	14	15	10	24	37	1.00000	1.00
4	1	1	MQ	394	31441	5	130	15	17	13	27	38	1.00000	1.00
5	1	2	EV	435	32815	5	115	15	16	10	23	42	0.80000	1.25
6	1	2	MQ	378	32101	5	110	15	16	12	28	31	0.80000	1.25
7	1	2	UA	409	33147	5	172	15	16	15	27	43	0.80000	1.25
8	0	2	MQ	405	34541	5	106	15	16	13	28	22	0.80000	0.00
9	1	2	DL	399	36825	5	213	15	17	8	26	32	0.80000	1.25
10	1	3	DL	415	35525	5	165	15	17	9	30	39	1.00000	1.00
11	1	3	EV	421	32083	5	126	15	17	4	31	27	1.00000	1.00
12	1	3	WN	431	34476	5	135	15	16	8	28	33	1.00000	1.00
13	1	4	WN	395	36185	5	124	15	17	4	26	32	0.83333	1.20
14	1	4	US	420	35736	5	127	15	17	13	29	40	0.83333	1.20
15	0	4	DL	424	33122	5	209	15	17	10	23	34	0.83333	0.00
16	1	4	WN	407	35670	6	117	15	16	6	28	34	0.83333	1.20
17	1	4	B6	437	34627	5	178	15	17	11	18	41	0.83333	1.20
18	1	4	B6	399	33949	5	133	15	17	11	16	40	0.83333	1.20
19	1	5	B6	422	34951	5	177	15	17	13	28	37	1.00000	1.00
20	1	5	B6	441	34868	5	131	16	16	11	24	44	1.00000	1.00

Now divide the data into training and testing data set based on the selection variable as seen in the above partial output (top 20 observations) fromTable 6-2. Where selection is equal to 1, assign all those observations to training data set and where selection is equal to 0, assign all those observations to testing data set. Training data set is used for model building and testing data set is used for testing the model.

```
/*Values of selected variable: 1 means selected for training  set, 0 means testing set*/
data libref.training;
    set libref.flight_delay_subset;
    if selected=1;
    proc print;
```

Table 6-6. *Partial Output of libref.training*

Obs	Selected	Arr_ Delay	Carrier	Airport_ Distance	Number_ of_ flights	Weather	Support_ Crew_ Available	Baggage_ loading_ time	Late_ Arrival_o	Cleaning_o	Fueling_o	Security_o	SelectionProb	Sampling Weight
1	1	0	MQ	395	29475	5	190	15	16	10	24	35	1.00000	1.00000
2	1	0	FL	418	30300	5	158	14	16	5	26	27	1.00000	1.00000
3	1	0	B6	376	29777	5	93	14	15	10	24	37	1.00000	1.00000
4	1	1	MQ	394	31441	5	130	15	17	13	27	38	1.00000	1.00000
5	1	2	EV	435	32815	5	115	15	16	10	23	42	0.80000	1.25000
6	1	2	MQ	378	32101	5	110	15	16	12	28	31	0.80000	1.25000
7	1	2	UA	409	33147	5	172	15	16	15	27	43	0.80000	1.25000
8	1	2	DL	399	36825	5	213	15	17	8	26	32	0.80000	1.25000
9	1	3	DL	415	35525	5	165	15	17	9	30	39	1.00000	1.00000
10	1	3	EV	421	32083	5	126	15	17	4	31	27	1.00000	1.00000

```
data libref.testing;
    set libref.flight_delay_subset;
    if selected=0;

    proc print;
```

Table 6-7. Partial output of libref.testing

Obs	Selected	Arr_ Delay	Carrier	Airport_ Distance	Number_ of_flights	Weather	Support_ Crew_ Available	Baggage_ loading_ time	Late_ Arrival_o	Cleaning_o	Fueling_o	Security_o	SelectionProb	Sampling Weight
1	0	2	MQ	405	34541	5	106	15	16	13	28	22	0.80000	0
2	0	4	DL	424	33122	5	209	15	17	10	23	34	0.83333	0
3	0	6	AA	413	33783	5	71	15	17	7	20	38	0.80000	0
4	0	7	DL	429	35068	5	187	15	16	13	27	35	0.75000	0
5	0	10	AA	439	37952	5	107	15	18	14	20	36	0.72727	0
6	0	10	MQ	451	36069	5	157	16	18	11	24	30	0.72727	0
7	0	10	WN	448	39131	5	75	16	18	11	31	27	0.72727	0
8	0	11	AA	412	37861	5	131	15	17	8	30	29	0.70000	0
9	0	11	WN	402	36633	5	154	16	18	8	27	32	0.70000	0
10	0	11	B6	404	39022	5	190	16	17	7	25	37	0.70000	0

In this section Training data set is used for model building.

In **program1** code **PROC REG** is used to build the linear regression model, **DATA** names a dataset to use for regression, in this case it is libref.training. **OUTTEST** outputs a data set (libref.Reg_P_Out) that contains parameter estimates of the model.

MODEL statement specifies dependent and independent variables in order to fit a model to the data. In this case, Arr_Delay is the dependent variable and Airport_Distance, Number_of_flights, Weather, Support_Crew_Available Baggage_loading_time, Late_Arrival_o, Cleaning_o, Fueling_o and Security_o are independent variables. **SELECTION** method is used to specify how independent variables are allowed to enter inside the analysis.There are various selection methods like stepwise, forward, backward. In this case the stepwise selection method is used. VIF prints the variance inflation factor statistics and is used to identify the multicollinearity. **DW** displays the Durbin-Watson statistics and detect the residual autocorrelation. **OUTPUT OUT** produces the output dataset libref.Reg_out_train. In this output data set **P=pred** displays the predicted values **R=Residual** displays the residual values (difference between Actual and Predicted values) **RSTUDENT=r1** displays the studentized residuals and detect the outliers **DFFITS=dffits** displays the DFFITS statistic **COOKD=cookd** displays the Cook's D statistic and detect the influential observations.

```
/*Program1: Building multiple linear regression model on training dataset*/

ODS GRAPHICS ON;
PROC REG DATA = libref.training OUTEST=libref.Reg_P_Out;
MODEL Arr_Delay = Airport_Distance Number_of_flights Weather Support_Crew_
Available Baggage_loading_time Late_Arrival_o Cleaning_o
     Fueling_o     Security_o/ SELECTION=stepwise VIF DW;
OUTPUT OUT = libref.Reg_out_train P=pred R=Residual RSTUDENT=r1
DFFITS=dffits COOKD=cookd;
RUN;
ODS GRAPHICS OFF;
```

The **Multiple linear regression output of program1** is split into several sections and each is discussed in the following section.

Part 1

The REG Procedure	
Model: MODEL1	
Dependent Variable: Arr_Delay	
Number of Observations Read	2597
Number of Observations Used	2597

Program 1 of part 1 displays the information that the number of observations read and used is 2597, hence there are no missing values present in the data.

Part 2

Stepwise Selection: Step 1

Variable Number_of_flights Entered: R-Square = 0.6839 and C(p) = 1964.627

Analysis of Variance

Source	DF	Sum of Squares	Mean Square	F Value	Pr > F
Model	1	1598587	1598587	5614.54	<.0001
Error	2595	738855	284.72262		
Corrected Total	2596	2337442			

Part 2 (*continued*)

Variable	Parameter Estimate	Standard Error	Type II SS	F Value	Pr > F
Intercept	-304.28441	5.00730	1051416	3692.77	<.0001
Number_of_ flights	0.00863	0.00011523	1598587	5614.54	<.0001

Bounds on condition number: 1, 1

In program 1 from part 2 to part 8, the stepwise regression is displayed and how all the variables are entered in the model one by one based on the F statistic significant at the SLENTRY = level. After all the variables are added, the variables that do not produce an F statistic significant at the SLSTAY = level are deleted. In this case the default value of the stepwise regression is considered for SLENTRY and SLSTAY. The stepwise procedure ends when none of the variables outside the model met the 0.1500 significance level for entry into the model (SLENTRY = level) and all variables left in the model are significant at 0.1500 level (SLSTAY = level). In part 2 Number_of_flights variable is entered as it met the criteria.

Part 3

Stepwise Selection: Step 2

Variable Baggage_loading_time Entered: R-Square = 0.7777 and C(p) = 613.9122

Analysis of Variance

Source	DF	Sum of Squares	Mean Square	F Value	Pr > F
Model	2	1817881	908940	4538.04	<.0001
Error	2594	519561	200.29342		
Corrected Total	2596	2337442			

Part 3 (*continued*)

Variable	Parameter Estimate	Standard Error	Type II SS	F Value	Pr > F
Intercept	-480.76832	6.78867	1004545	5015.37	<.0001
Number_of_flights	0.00563	0.00013255	361687	1805.78	<.0001
Baggage_loading_time	18.05676	0.54571	219294	1094.86	<.0001

Bounds on condition number: 1.881, 7.5239

In part 3 Number_of_flights and Baggage_loading_time variables are entered as they met the criteria.

Part 4

Stepwise Selection: Step 3

Variable Late_Arrival_o Entered: R-Square = 0.8035 and C(p) = 243.9176

Analysis of Variance

Source	DF	Sum of Squares	Mean Square	F Value	Pr > F
Model	3	1878186	626062	3534.80	<.0001
Error	2593	459256	177.11362		
Corrected Total	2596	2337442			

Part 4 (*continued*)

Variable	Parameter Estimate	Standard Error	Type II SS	F Value	Pr > F
Intercept	-544.15127	7.24923	997947	5634.50	<.0001
Number_of_flights	0.00481	0.00013231	234475	1323.87	<.0001
Baggage_loading_time	15.33640	0.53392	146134	825.09	<.0001
Late_Arrival_o	7.74204	0.41957	60306	340.49	<.0001

Bounds on condition number: 2.1193, 17.437

In part 4 Number_of_flights, Baggage_loading_time and Late_Arrival_o variables are entered as they met the criteria.

Part 5

Stepwise Selection: Step 4
Variable Airport_Distance Entered: R-Square = 0.8120 and C(p) = 123.0486
Analysis of Variance

Source	DF	Sum of Squares	Mean Square	F Value	Pr > F
Model	4	1898105	474526	2799.61	<.0001
Error	2592	439337	169.49723		
Corrected Total	2596	2337442			

Part 5 (*continued*)

Variable	Parameter Estimate	Standard Error	Type II SS	F Value	Pr > F
Intercept	-591.61391	8.33431	854084	5038.92	<.0001
Airport_Distance	0.18359	0.01694	19919	117.52	<.0001
Number_of_flights	0.00458	0.00013123	206455	1218.05	<.0001
Baggage_loading_time	14.30547	0.53090	123068	726.08	<.0001
Late_Arrival_o	7.41858	0.41153	55080	324.96	<.0001

Bounds on condition number: 2.1784, 28.895

In part 5 Airport_Distance, Number_of_flights, Baggage_loading_time and Late_Arrival_o variables are entered as they met the criteria.

Part 6

Stepwise Selection: Step 5
Variable Weather Entered: R-Square = 0.8163 and C(p) = 64.0653
Analysis of Variance

Source	DF	Sum of Squares	Mean Square	F Value	Pr > F
Model	5	1907991	381598	2302.29	<.0001
Error	2591	429451	165.74703		
Corrected Total	2596	2337442			

Part 6 (*continued*)

Variable	Parameter Estimate	Standard Error	Type II SS	F Value	Pr > F
Intercept	-599.03362	8.29740	863902	5212.17	<.0001
Airport_Distance	0.18087	0.01675	19324	116.59	<.0001
Number_of_flights	0.00452	0.00012998	200612	1210.35	<.0001
Weather	4.25731	0.55124	9886.25645	59.65	<.0001
Baggage_loading_time	13.79740	0.52910	112712	680.02	<.0001
Late_Arrival_o	7.25571	0.40750	52548	317.03	<.0001

Bounds on condition number: 2.1856, 41.812

In part 6 Airport_Distance, Number_of_flights, Weather, Baggage_loading_time and Late_Arrival_o variables are entered as they met the criteria.

Part 7

Stepwise Selection: Step 6
Variable Support_Crew_Available Entered: R-Square = 0.8203 and C(p) = 7.9886
Analysis of Variance

Source	DF	Sum of Squares	Mean Square	F Value	Pr > F
Model	6	1917406	319568	1970.50	<.0001
Error	2590	420036	162.17588		
Corrected Total	2596	2337442			

Part 7 (*continued*)

Variable	Parameter Estimate	Standard Error	Type II SS	F Value	Pr > F
Intercept	-578.28912	8.64731	725294	4472.27	<.0001
Airport_Distance	0.17716	0.01658	18523	114.22	<.0001
Number_of_flights	0.00442	0.00012931	189262	1167.02	<.0001
Weather	4.19333	0.54534	9589.07148	59.13	<.0001
Support_Crew_Available	-0.04843	0.00636	9415.04763	58.05	<.0001
Baggage_loading_time	13.41806	0.52573	105643	651.41	<.0001
Late_Arrival_o	7.05988	0.40390	49548	305.52	<.0001

Bounds on condition number: 2.2106, 57.236

In part 7 Airport_Distance, Number_of_flights, Weather, Baggage_loading_time Support_Crew_Available and Late_Arrival_o variables are entered as they met the criteria.

Part 8

Stepwise Selection: Step 7					
Variable Fueling_o Entered: R-Square = 0.8205 and C(p) = 7.3028					
Analysis of Variance					
Source	DF	Sum of Squares	Mean Square	F Value	Pr > F
Model	7	1917842	273977	1690.48	<.0001
Error	2589	419600	162.07034		
Corrected Total	2596	2337442			

Part 8 (*continued*)

Variable	Parameter Estimate	Standard Error	Type II SS	F Value	Pr > F
Intercept	-575.24770	8.84142	686072	4233.17	<.0001
Airport_Distance	0.17577	0.01659	18186	112.21	<.0001
Number_of_flights	0.00441	0.00012937	188210	1161.29	<.0001
Weather	4.22902	0.54559	9737.46361	60.08	<.0001

(*continued*)

Variable	Parameter Estimate	Standard Error	Type II SS	F Value	Pr > F
Support_Crew_ Available	-0.04832	0.00635	9372.00589	57.83	<.0001
Baggage_loading_ time	13.45121	0.52595	106009	654.09	<.0001
Late_Arrival_o	7.06580	0.40379	49627	306.21	<.0001
Fueling_o	-0.11710	0.07144	435.39740	2.69	0.1013

Bounds on condition number: 2.2143, 73.908
All variables left in the model are significant at the 0.1500 level.
No other variable met the 0.1500 significance level for entry into the model.

In part 7 Airport_Distance, Number_of_flights, Weather, Baggage_loading_time Support_Crew_Available, Late_Arrival_o and Fueling_o variables are entered as they met the criteria.

Part 9

Summary of Stepwise Selection

Step	Variable Entered	Variable Removed	Number Vars In	Partial R-Square	Model R-Square	C(p)	F Value	Pr > F
1	Number_of_ flights		1	0.6839	0.6839	1964.63	5614.54	<.0001
2	Baggage_ loading_time		2	0.0938	0.7777	613.912	1094.86	<.0001
3	Late_Arrival_o		3	0.0258	0.8035	243.918	340.49	<.0001
4	Airport_ Distance		4	0.0085	0.8120	123.049	117.52	<.0001
5	Weather		5	0.0042	0.8163	64.0653	59.65	<.0001
6	Support_ Crew_Available		6	0.0040	0.8203	7.9886	58.05	<.0001
7	Fueling_o		7	0.0002	0.8205	7.3028	2.69	0.1013

In part 9, all the variables that met the criteria are displayed and entered in the model such as Number_of_flights, Baggage_loading_time, Late_Arrival_o, Airport_Distance, Weather, Support_Crew_Available and Fueling_o.

Part 10

The REG Procedure	
Model: MODEL1	
Dependent Variable: Arr_Delay	
Number of Observations Read	2597
Number of Observations Used	2597

Program 1 of part 10 displays the information that the number of observations read and used is 2597, hence there are no missing values in the data.

Part 11

Analysis of Variance					
Source	DF	Sum of Squares	Mean Square	F Value	Pr > F
Model	7	1917842	273977	1690.48	<.0001
Error	2589	419600	162.07034		
Corrected Total	2596	2337442			

Program 1 of part 11 displays the Analysis of variance table. The F value of 1690.48 with an associated p - value less than 0.0001 indicates a significant relationship between the dependent variable Arr_Delay and at least one of the independent variables in the model.

Part 12

Root MSE	12.73069	R-Square	0.8205
Dependent Mean	70.09241	Adj R-Sq	0.8200
Coeff Var	18.16271		

Program 1 of part 12 displays R-square (0.8205), Adj R-square (0.8200) and Root MSE (12.73069). In this case R-square value indicates that the 82% of the variation of Arr_delay is explained by the independent variables in the model.

Part 13

Parameter Estimates

Variable	DF	Parameter Estimate	Standard Error	t Value	Pr > |t|	Variance Inflation
Intercept	1	-575.24770	8.84142	-65.06	<.0001	0
Airport_ Distance	1	0.17577	0.01659	10.59	<.0001	1.28107
Number_of_ flights	1	0.00441	0.00012937	34.08	<.0001	2.21428
Weather	1	4.22902	0.54559	7.75	<.0001	1.09538
Support_Crew_ Available	1	-0.04832	0.00635	-7.60	<.0001	1.12443
Baggage_ loading_time	1	13.45121	0.52595	25.58	<.0001	2.15928
Late_Arrival_o	1	7.06580	0.40379	17.50	<.0001	1.67702
Fueling_o	1	-0.11710	0.07144	-1.64	0.1013	1.00678

Program 1 of part 13 displays the "Parameter Estimates" table. This table lists the degrees of freedom (DF), the parameter estimates, the standard error of the estimates, the t values, the associated probabilities (p-values) and the VIF. All the variables except Fueling_o p-values str less than 0.05, thus all the variables are significant except Fueling_o. VIF greater than 5 is considered as the collinearity case, in this case none of the variable VIFs are greater than 5, hence no collinearity.

Part 14

The REG Procedure	
Model: MODEL1	
Dependent Variable: Arr_Delay	
Durbin-Watson D	1.323
Number of Observations	2597
1st Order Autocorrelation	0.335

Program 1 of part 14 displays the Durbin-Watson D value as 1.323 and detects the residual autocorrelation.

Part 15

The REG Procedure

Model: MODEL1

Dependent Variable: Arr_Delay

Fit Diagnostics plots for Arr_Delay is displayed in Figure 6-12.

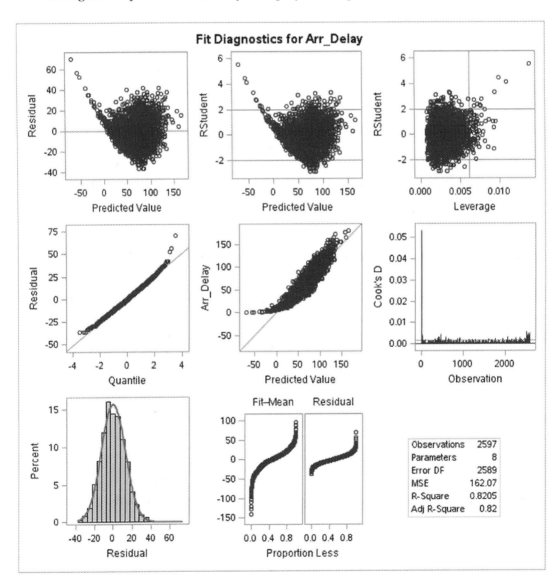

Figure 6-12. *Fit Diagnostics plot for Arr_Delay*

Program 1 of part 15 displays the diagnostic plots. These plots are helpful in detecting any violations in assumptions of the linear regression model. The Residuals versus predicted values plot does not show any pattern, if there are any patterns like cone, sphere, etc., then it indicates lack of model fit and unequal variances. In Q-Q plot a linear trend is seen only in a slight deviation at the tail, suggesting that the normality assumption is satisfied. The histogram shows the normal distribution. In Studentized residual versus Leverage graph, some of the outliers and influential observations are seen outside the reference lines and Cook's D graph also shows influential observations.

Residual by Regressors plots for Arr_Delay is displayed in Figure 6-13.

Part 16

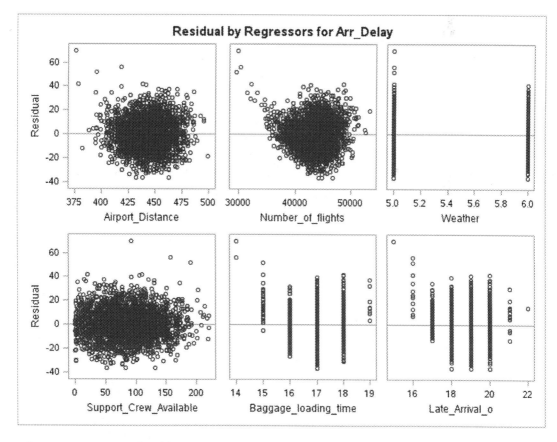

Figure 6-13.   *Residual by Regressors for Arr_Delay plot*

Part 17

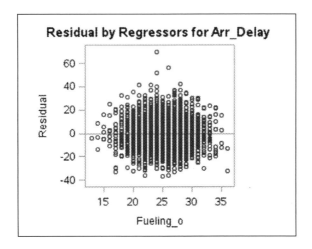

Figure 6-14. *Residual by Regressors for Arr_Delay plot*

Program 1 of part 16 and 17 displays the residual plots

PROC SCORE is used to score a libref.testing dataset. In this case SCORE=libref.
Reg_P_Out the parameter estimates are output to the libref.Reg_P_Out dataset and is
used as scoring coefficients. PREDICT OUT= libref.R_Score_Pred data set is produced
and the VAR statement in the PROC SCORE includes only the independent variables
from the model in PROC REG and the Arr_Delay variable contains the predicted values.

```
/*Score procedure on testing data*/
PROC SCORE DATA= libref.testing SCORE=libref.Reg_P_Out TYPE = parms PREDICT
OUT= libref.R_Score_Pred;
VAR Airport_Distance Number_of_flights Weather Support_Crew_Available
Baggage_loading_time Late_Arrival_o  Cleaning_o Fueling_o Security_o;
run;

proc print data= libref.R_Score_Pred;
title 'Predicted Scores for Regression on Testing Data';
run;
```

In Table 6-8 libref.R_Score_Pred data displays the partial output of predicted scores
for regression on testing data.

Table 6-8. *Partial Output of Predicted Scores for Regression on Testing Data*

Predicted Scores for Regression on Testing Data

Obs	Selected	Arr_Delay	Carrier	Airport_ Distance	Number_ of_flights	Weather	Support_ Crew_ Available	Baggage_ loading_ time	Late_ Arrival_o	Cleaning_o	Fueling_o	Security_o	SelectionProb	Sampling Weight	MODEL1
1	0	2	MQ	405	34541	5	106	15	16	13	28	22	0.80000	0	-24.215
2	0	4	DL	424	33122	5	209	15	17	10	23	34	0.83333	0	-24.457
3	0	6	AA	413	33783	5	71	15	17	7	20	38	0.80000	0	-16.457
4	0	7	DL	429	35068	5	187	15	16	13	27	35	0.75000	0	-21.470
5	0	10	AA	439	37952	5	107	15	18	14	20	36	0.72727	0	11.819
6	0	10	MQ	451	36069	5	157	16	18	11	24	30	0.72727	0	16.194
7	0	10	WN	448	39131	5	75	16	18	11	31	27	0.72727	0	32.308
8	0	11	AA	412	37861	5	131	15	17	8	30	29	0.70000	0	-2.724
9	0	11	WN	402	36633	5	154	16	18	8	27	32	0.70000	0	9.861
10	0	11	B6	404	39022	5	190	16	17	7	25	37	0.70000	0	12.174
11	0	12	AA	424	38695	5	81	16	18	14	21	36	0.76923	0	27.049
12	0	12	AA	429	39822	5	111	16	18	8	24	31	0.76923	0	31.095
13	0	12	EV	402	37707	5	141	16	18	15	25	42	0.76923	0	15.458
14	0	13	MQ	430	37899	6	134	15	18	12	24	35	0.70000	0	12.460
15	0	13	EV	415	37476	5	116	16	18	17	26	31	0.70000	0	17.816

(continued)

335

Table 6-8. (continued)

Predicted Scores for Regression on Testing Data

Obs	Selected	Arr_Delay	Carrier	Airport_Distance	Number_of_flights	Weather	Support_Crew_Available	Baggage_loading_time	Late_Arrival_o	Cleaning_o	Fueling_o	Security_o	SelectionProb	Sampling Weight	MODEL1
16	0	13	B6	404	37657	5	95	16	18	7	32	35	0.70000	0	16.993
17	0	14	MQ	422	36944	5	143	16	18	16	21	42	0.76923	0	15.982
18	0	14	UA	426	37577	5	110	16	17	12	36	32	0.76923	0	12.248
19	0	14	B6	419	39764	5	168	16	18	8	26	39	0.76923	0	26.093
20	0	15	UA	394	37974	5	174	16	18	7	28	32	0.75000	0	13.283
21	0	15	DL	443	39137	5	106	16	18	9	25	44	0.75000	0	30.661
22	0	15	DL	428	40074	5	139	16	18	12	23	29	0.75000	0	30.795
23	0	16	UA	412	36038	5	154	16	18	11	23	40	0.70000	0	9.464
24	0	16	DL	444	36458	5	95	16	18	12	27	39	0.70000	0	19.323
25	0	16	DL	421	38514	5	98	15	18	12	28	48	0.70000	0	10.631
26	0	17	AA	400	38298	6	41	16	18	10	21	20	0.71429	0	27.242
27	0	17	B6	450	39244	5	183	16	18	11	32	37	0.71429	0	27.822
28	0	18	MQ	448	38237	5	56	16	17	11	25	45	0.72727	0	22.922
29	0	18	VX	437	37199	5	143	16	18	16	28	43	0.72727	0	18.923
30	0	18	AA	419	38672	5	137	16	18	5	31	41	0.72727	0	22.192

PROC SCORE is used to score a libref.testing dataset. In this case SCORE=libref. Reg_P_Out the parameter estimates are output to the libref.Reg_P_Out dataset and is used as scoring coefficients. RESIDUAL OUT= libref. R_Score_R data set is produced and the VAR statement in the PROC SCORE includes dependent and independent variables from the model in PROC REG and the Arr_Delay variable contains the positive residuals (Actual – Predict). If the RESIDUAL option is not stated in the above code, then Arr_ Delay variable contains the negative residuals(Predict- Actual).

```
/*Residual Scores from the Data Set Created by PROC SCORE*/

PROC SCORE DATA= libref.testing SCORE=libref.Reg_P_Out RESIDUAL OUT=
libref.R_Score_R TYPE = parms;
VAR Arr_Delay  Airport_Distance    Number_of_flights Weather Support_
Crew_Available Baggage_loading_time Late_Arrival_o  Cleaning_o
Fueling_o  Security_o;
run;

proc print data = libref.R_Score_R;
title 'Residual Scores for Regression on Testing Data';
run;
```

In Table 6-9, libref.R_Score_R data displays the partial output of residual scores for regression on testing data.

Table 6-9. Partial Output of Residual Scores for Regression on Testing Fata

Residual Scores for Regression on Testing Data

Obs	Selected	Arr_Delay	Carrier	Airport_Distance	Number_of_flights	Weather	Support_Crew_Available time	Baggage_loading_time	Late_Arrival_o	Cleaning_o	Fueling_o	Security_o	SelectionProb	Sampling Weight	MODEL1
1	0	2	MQ	405	34541	5	106	15	16	13	28	22	0.80000	0	26.2151
2	0	4	DL	424	33122	5	209	15	17	10	23	34	0.83333	0	28.4570
3	0	6	AA	413	33783	5	71	15	17	7	20	38	0.80000	0	22.4569
4	0	7	DL	429	35068	5	187	15	16	13	27	35	0.75000	0	28.4701
5	0	10	AA	439	37952	5	107	15	18	14	20	36	0.72727	0	-1.8192
6	0	10	MQ	451	36069	5	157	16	18	11	24	30	0.72727	0	-6.1937
7	0	10	WN	448	39131	5	75	16	18	11	31	27	0.72727	0	-22.3083
8	0	11	AA	412	37861	5	131	15	17	8	30	29	0.70000	0	13.7243
9	0	11	WN	402	36633	5	154	16	18	8	27	32	0.70000	0	1.1389
10	0	11	B6	404	39022	5	190	16	17	7	25	37	0.70000	0	-1.1739
11	0	12	AA	424	38695	5	81	16	18	14	21	36	0.76923	0	-15.0487
12	0	12	AA	429	39822	5	111	16	18	8	24	31	0.76923	0	-19.0952
13	0	12	EV	402	37707	5	141	16	18	15	25	42	0.76923	0	-3.4584
14	0	13	MQ	430	37899	6	134	15	18	12	24	35	0.70000	0	0.5404

15	0	13	EV	415	37476	5	116	16	18	17	26	31	0.70000	0	-4.8159
16	0	13	B6	404	37657	5	95	16	18	7	32	35	0.70000	0	-3.9925
17	0	14	MQ	422	36944	5	143	16	18	16	21	42	0.76923	0	-1.9817
18	0	14	UA	426	37577	5	110	16	17	12	36	32	0.76923	0	1.7522
19	0	14	B6	419	39764	5	168	16	18	8	26	39	0.76923	0	-12.0934
20	0	15	UA	394	37974	5	174	16	18	7	28	32	0.75000	0	1.7165
21	0	15	DL	443	39137	5	106	16	18	9	25	44	0.75000	0	-15.6606
22	0	15	DL	428	40074	5	139	16	18	12	23	29	0.75000	0	-15.7946
23	0	16	UA	412	36038	5	154	16	18	11	23	40	0.70000	0	6.5360
24	0	16	DL	444	36458	5	95	16	18	12	27	39	0.70000	0	-3.3228
25	0	16	DL	421	38514	5	98	15	18	12	28	48	0.70000	0	5.3689
26	0	17	AA	400	38298	6	41	16	18	10	21	20	0.71429	0	-10.2418
27	0	17	B6	450	39244	5	183	16	18	11	32	37	0.71429	0	-10.8224
28	0	18	MQ	448	38237	5	56	16	17	11	25	45	0.72727	0	-4.9218
29	0	18	VX	437	37199	5	143	16	18	16	28	43	0.72727	0	-0.9228
30	0	18	AA	419	38672	5	137	16	18	5	31	41	0.72727	0	-4.1915

Summary

In this chapter we have learned about the multiple linear regression model. We also have discussed various characteristics, features, and assumptions of the model. Practical application of this model in a real-life scenario was demonstrated by help of a case study to predict flight arrival delay, which is a big challenge for the airline industry. From model development stage to execution, visualization, and result interpretations were also performed by using both R and SAS Studio.

References

1. Association, I. A. T., Strong Airline Profitability Continues in 2018. `http://www.iata.org/pressroom/pr/Pages/2017-12-05-01.aspx`, 2017.

2. Association, I. A. T., IATA Forecasts Passenger Demand to Double Over 20 Years. `http://www.iata.org/pressroom/pr/Pages/2016-10-18-02.aspx`.

3. Falcus, M. Top 10 World's Largest Airlines 2016.

4. Awards, S. W. A. Qatar Airways is announced as the World's Best Airline at the 2017 World Airline Awards.

5. Cleofe Maceda, Skytrax names world's best airlines for 2017 Three carriers from the Gulf including Etihad, Emirates land in top ten, honoured during ceremony in Paris. Gulf News: `http://gulfnews.com/business/aviation/skytrax-names-world-s-best-airlines-for-2017-1.2047274`.

6. Symonds, T., Brussels attacks: Airport security under the spotlight again. BBC: `http://www.bbc.com/news/world-35873989`, 2016.

7. BBC, Missing Malaysia plane MH370: What we know. `http://www.bbc.com/news/world-asia-26503141`, 2017.

8. BBC. MH17 Ukraine plane crash: What we know 2016.

9. Association, I. A. T. Global Aviation Data Management (GADM).

10. international, A. c. Airport Carbon Accreditation; `http://www.airportcarbonaccredited.org/library/annual-reports.html`, 2017; p 96.

11. organization, I. c. a. Air Carbon Emissions Management; 2015; p 4.

12. Georghiades, C. P. H. a. E. In An Analysis of Consumer Search and Buying Behaviour in the US Airline Industry using Big Data, ENTER 2015, 2015; p 5.

13. Royo-Vela, M.; Martinez-Garcia, E., A Segmentation Analysis and segments profile of budget air travelers. 2010; p 241-259.

14. Siau, K.; E, A. In An Approach to Sentiment Analysis – The Case of Airline Quality Rating, 18th Pacific Asia Conference on Information Systems · PACIS 2014, Chengdu, China, Chengdu, China, 2014; p 12.

15. Ravi, L.; Vairavasundaram, S., A Collaborative Location Based Travel Recommendation System through Enhanced Rating Prediction for the Group of Users. Computational Intelligence and Neuroscience 2016, 2016, 28.

16. Laskowski, N. NASA uses text analytics to bolster aviation safety Search Business Analytics [Online], 2012.

17. Administration, N. A. a. S. NASA Partners with Radiometrics Corporation to Improve Weather Forecasting and Flight Safety.

18. Green, S. In A Study of U.S. Inflight Icing Accidents and Incidents, 1978 to 2002, 44th AIAA Aerospace Sciences Meeting and Exhibit, Reno, Nevada, Reno, Nevada, 2006.

19. Thompson, G.; Politovich, M. K.; Rasmussen, R. M., A Numerical Weather Model's Ability to Predict Characteristics of Aircraft Icing Environments. Weather and Forecasting 2016, 32 (1), 207-221.

20. Wald, M. L., The crash of flight 4184: The overview; Flight Recorders Found At Indiana Crash Scene. The New York Times 1994.

21. Board, N. T. S. In-flight Icing Encounter and Loss of Control Simmons Airlines, d.b.a. American Eagle Flight 4184 Avions de Transport Regional (ATR) Model 72-212, N401AM; https://www.ntsb.gov/investigations/AccidentReports/Pages/AAR9601.aspx, 1996.

22. Association, I. A. T. IATA Industry Fraud Prevention, p. 3.

23. Cooperation, E. U. A. f. L. E., 195 Individuals detained as a result of global crackdown on airlines ticket fraud. https://www.europol.europa.eu/newsroom/news/195-individuals-detained-result-of-global-crackdown-airline-ticket-fraud, 2017.

24. statistics, B. o. t. Understanding the Reporting of Causes of Flight Delays and Cancellations.

25. Transportation, U. S. D. o., Air Travel Consumer Reports. https://www.transportation.gov/airconsumer/air-travel-consumer-reports, 2018.

26. Department, U. S.; Transportation, o. Air Travel Consumer Report; https://www.transportation.gov/sites/dot.gov/files/docs/resources/individuals/aviation-consumer-protection/30355 6/2018januaryatcr.pdf, 2018; p 46.

27. Administration, N. A. a. S. NASA Aircraft Arrival Technology Gets Big Test in 2017 2016.

28. Uyanık, G. K.; Güler, N., A Study on Multiple Linear Regression Analysis. Procedia - Social and Behavioral Sciences 2013, 106, 234-240.

29. Tranmer, M.; Elliot, M. Multiple Linear Regression; Cathie Marsh Center of Census and Survey Research: http://hummedia.manchester.ac.uk/institutes/cmist/archive-publications/working-papers/2008/2008-19-multiple-linear-regression.pdf, 2008; p 47.

30. Allison, P. D., Multiple Regression: A Primer. SAGE Publications: 1999.

31. Alexopoulos, E. C., Introduction to Multivariate Regression Analysis. Hippokratia 2010, 14 (Suppl 1), 23-28.

32. Blatná, D., Outliers in Regression. University of Economics Prague: 2006; p 6.

33. Jacoby, W. G., Regression III: Advanced Methods. Michigan State University: `http://polisci.msu.edu/jacoby/icpsr/regress3/lectures/week3/11.Outliers.pdf`.

34. Freund, R. J.; Wilson, W. J., Regression Analysis: Statistical Modeling of a Response Variable. Academic Press: 1998.

CHAPTER 7

FMCG Case Study

The Fast Moving Consumer Good (FMCG) industry primarily deals with the production, distribution, and marketing of consumer-packaged goods. As the definition suggests, FMCG goods are also popularly called CPG (Consumer-packaged goods). The Fast Moving Consumer Goods (FMCG) are those products that are consumed by the consumers at regular intervals. In the FMCG industry, the products have turnover quickly at a relatively low cost.[1] This industry operates as a high volume business and less margin. This segment can be classified into premium segments and popular segments. The premium segment consists of the consumers from higher income strata and is not sensitive to the price but is more quality and brand conscious. On one hand, popular segments target the consumers from medium to low income strata who are not brand conscious and are very sensitive to the price. The products sold to the popular segments are lower in prices as compared to the premium segments. The industry is comprised of consumers and non-durable products and is broadly classified under three categories displayed in Figure 7-1.

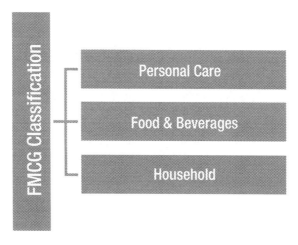

Figure 7-1. *FMCG Classification*

© Deepti Gupta 2018
D. Gupta, *Applied Analytics through Case Studies Using SAS and R,*
https://doi.org/10.1007/978-1-4842-3525-6_7

The FMCG industry is a substantial global industry. According to ASSOCHAM-Tech Si report, India's FMCG market is expected to reach USD 104 billion by 2020.[2] The term "fast moving" means that the FMCG product's shelf life is very short and the bulk of products fall under non-durables. Some of the most renowned FMCG companies in the global market are Pepsi, Coca-Cola, Kellogg's, Kraft, Diageo, Heineken, Nestle, Unilever, Procter & Gamble, and Johnson & Johnson, etc.[3] From the consumer's perspective the main characteristics of FMCG industry are low product prices, frequent purchases, fast consumption, and short shelf life of products. From the marketer's perspective, it is high volume, low margin, high stock turnover, and a widespread distribution network.

In this chapter the key applications of analytics, which have redefined the FMCG industry are discussed. Introduction to the RFM model and K-means clustering is provided. A detailed case study on customer segmentation based on their purchase history is presented in both R and SAS Studio.

Application of Analytics in FMCG Industry

Success in the FMCG industry relies on sales volumes and speed due to low prices and the non-durable nature of products. In terms of data, this means vast amounts of data needs to be analyzed and models built should have quick predictability and ability to track variables that are changing fast and influencing sales. FMCG data is collected from different sources like sales record, dealers, customers, and salesmen; and nontraditional sources like weather forecasting, traffic patterns, etc. The FMCG industry is facing challenges in terms of managing the data produced on a daily basis in different formats like texts, images, etc., originating from disparate systems. Big data technologies are being implemented for storage, accessibility, security, and overall management of this vast amount of data. This vast amount of data offers an opportunity to employ predictive analytics to address the complex issues faced by the FMCG business by effectively withdrawing the relevant insights and translating it into usable information to support all key business decisions in the industry. Predictive Analytics is playing an important role in redefining FMCG industry in a holistic sense with applications ranging from customer experience & engagement, sales & marketing, logistics management, and markdown optimization. An overview of some of these key applications is provided in this section.

Customer Experience & Engagement

In today's digital world, customers not only expect a satisfactory personalized experience during their sales journey but also anticipate customer engagement.[4] Traditional analytics used the "mass segment" approach but modern analytics is using "segment-of-one" marketing approach. Each customer has different desires, different needs; therefore, almost all top FMCG companies are using advance analytics in order to trim a segment down to an individual. This "segment-of-one" marketing approach helps all these companies to track and understand the behavior of the customers in a better way and convert them into loyal, high-valued customers at more profitable rates.

Combining predictive analytics with social media analytics offers a powerful platform for identifying the trends, patterns, correlations, and sentiments.[5] It can provide effective insight to the FMCG companies as to what the customers think about their products and services; what are their expectations, needs, wants, behaviors, and intentions. These inputs help the companies in analyzing whether they are making the right investment in their products and services, marketing strategies, employees, and customer service. Knowing more about the individual customer gives a better opportunity to understand the customer's perception about their brand with that of the main competitors and create a better data-driven brand strategy. They can then provide relevant offers to the customers by identifying their shopping behaviors and habits by doing customer segmentation across digital and traditional channels. They can engage with the customers through the right channel with right offers at the right time. Churn prediction helps companies to know which customers have a high likelihood to churn and create effective strategies to retain them. For FMCG companies, in this digital economy, personalized experience and customer engagement are vital for their success.

Sales and Marketing

In the modern FMCG industry, the number of sales transactions is huge. Sales data plays an essential role in increasing the sales, which is the primary objective of all the FMCG companies. Accurate sales forecasting is very important for all the FMCG companies because of the short shelf life of their products and uncertainty and fluctuations in consumer demands.[6] Various well-known time series forecasting methods like Moving Average, Exponential smoothing, ARIMA, and ANN are used by the FMCG companies for forecasting short shelf-lives of food products.[7] Considering only the sales figures for the sales forecast is not the appropriate method as it can result the inaccurate in sales forecasts; therefore, management lacks the confidence in these sales forecast numbers for making the business decisions. In today's big data world by using advanced technology, the data can be collected from different sources easily (sales, marketing, demographics, digital, weather, etc.). All this data is merged easily and effective insights can be withdrawn from the data, which improves the quality of the sales forecast in FMCG companies. For example, merging marketing data with the sales data leads to more accurate marketing predictions; and it is also possible that we can fetch the information from vast data about the products, which products should be promoted in the month, type of campaign that will fetch more profit for the particular products, targeting the most profitable customers, and many more. Predictive Analytics is leveraged to build marketing mix models, which help in estimating the potential value of all marketing inputs and help in discovering better marketing investments for higher revenue. The marketing mix model helps in guiding how to spend, thus reducing ineffective spending.[8] It creates transparency across multiple marketing levers, including drivers of the performance, the return on investment of different marketing levers, and impact on customer price perception. Marketing mix model optimizes marketing ROI by using scenario analysis and optimization models.

Logistics Management

In FMCG companies effective logistics is highly demanded and plays key roles as the suppliers have to manage the large volumes and the fast-changing rates with the distributors. In logistics management, analysis is mainly focused on optimizing delivery, shipments, and warehousing performances.[9] Inventory optimization is about having sufficient stock so the customers receive their goods on time and the holding and ordering costs must be low. Route optimization is not only saving the cost for FMCG companies but also increasing the business revenue by selling efficiently and selling more. Route is the basic building unit for sales and distribution in FMCG companies, which aims in smoothing the process of delivery in minimum time, minimum distance, and minimum cost of fuels. For example, it involves how each route is progressing, number of truck drivers on the road, number of the deliveries each driver is doing, where their last stop was, temperature, transport duration, and more. Especially applying intelligent logistics in perishable segment reduce the losses due to temperature and transport conditions.[10] The advance algorithms help the companies in suggesting the shortest and ideal routes, predicting the traffic at the peak hours, predicting on-time delivery, suggesting the right zones, and proper clusters that optimize market proximity from distribution centers. Reverse logistics is another important factor in having effective logistics management in FMCG companies.[11] Without effective monitoring and reverse logistics management the companies could be losing millions of dollars in potential revenue. Properly managing the reverse logistics or a strong logistics reverse program helps the FMCG companies not only in reducing the cost but also increasing the revenue. It can also help in maintaining consumer loyalty and satisfaction. Figure 7-2 displays the reverse logistics cycle.

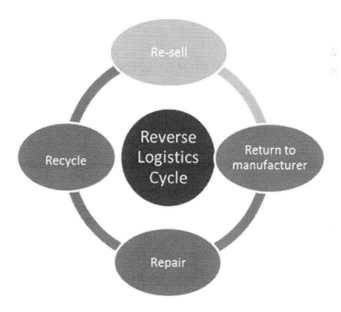

Figure 7-2. *Reverse logistics cycle*

Markdown Optimization

Markdown optimization plays an important role in FMCG companies for making strategic decisions. A carefully implemented markdown optimization results in more sell–through rates, increased margins, and higher revenues.[12] FMCG companies face challenges in deciding the optimal pricing for their products in different seasons. During the peak seasons they need to fix the right price for the product and create effective strategies for marketing, in order to lure more customers and generate higher sales and profit. While in the off-season, they need to decrease the prices of the same product so the stocks are cleared before the trend is outdated and this process is known as markdown. In markdown optimization, various factors that impact the demand like behavior of customer, market trend, purchase history of customers, pricing, and more are analyzed in order to fix the right price for the product that will not only trigger the increase in the demand for the product but also increase the profit margin. Customers are highly sensitive to the pricing so the data scientist considers behavior of customer, trends and cost, etc., for predicting customer behavior with the change in price. Applying analytics to perform markdown optimization helps the companies in providing optimal discounts to the products at the right time across the multiple channels, resulting in increased revenue.

Case Study: Customer Segmentation with RFM Model and K-means Clustering

Fast Moving Consumer goods (FMCG) are a multimillion-dollar sector that tends to be high-volume and low-cost items. In this industry, the items quickly leave the supermarket shelves. The top FMCG companies are producing the items that are in highest demand by the customers and simultaneously develop customer loyalty and trust toward their brands. According to the **Pareto Principle** 80% of the revenue comes from t20% of loyal customers.[13] From a marketing perspective it is essential to understand the customer's needs in a better way than their competitor and customize their products and items according to their needs. It is also essential in targeting the prospects who resemble the best customers with effective marketing strategies. Customer relationship management is the strategy used by all the firms in today's competitive business in order to acquire and retain their customers. Analysts are using different methods to analyze customer value in a better way. The RFM method is one of the most common methods used by all the marketers to segment and identify customer value in the organization. The RFM method is based on recency (when), frequency (how often), and monetary (how much money is spent) for particular product or service. The RFM model can be used in conjunction with other predictive models like K-means clustering, logistic regression, decision tree, etc., in order to provide better insights about the customer. In this case study, we used the results of RFM method as inputs for K-means clustering for determining customer loyalty.

Overview of RFM Model

Customer relationship management is critical of all the FMCG companies and for that, there is need to identify the valuable customers. For this a simple proven and highly effective technique, a marketing model called RFM (Recency, Frequency, and Monetary) is used. In this model, the customer segmentation is based on, how recently, how frequently, and how much money is spent by a particular customer.[14]

   RFM Analysis is used by marketers to identify the customers who have a high likelihood to respond to the marketing campaigns, promotions, etc. It also helps in targeting for future personalized service. Measuring the customer's value based on a single parameter is not enough. For example, most of us think that the customers who spend the most are the most valuable and loyal customers but this is not the case. What if the customer purchased your product only once and purchased it a long time ago; then do you think still those customers are the most valuable? Probably not. Therefore, the RFM model is used to know the customer base and their lifetime value accurately. The RFM model uses the three attributes of recency, frequency, and monetary in order to rank their customers. By using the recency, frequency and monetary scores, customer segmentation into different groups is done. Customers are considered as high valued or profitable customers when their RFM scores are highest. Figure 7-3 displays the profitable customers based on the highest RFM scores (555 OR 333).

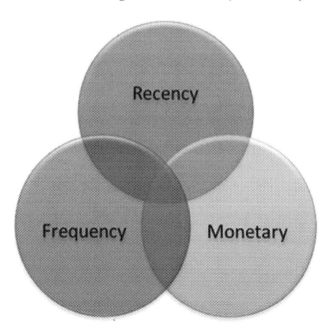

Figure 7-3. *Profitable customers based on highest RFM scores (555 or333)*

In this section, variables that make up the RFM score are discussed in more detail.

1) Recency: It displays the information about that how recently the customer made a purchase order with your business. According to the RFM model the customer who has purchased recently from your store has more likelihood to respond to any new initiative in terms of campaigns, promotions, etc., as compared to the customer who have has not purchased from you for a long time.

2) Frequency: How often the customer makes the purchase from your business. A customer who has purchased frequently or more often from your store has more likelihood to purchase again from your store as compared to the infrequent customer. Eventually the frequency of the purchase statistics is used to predict the likelihood of future purchases from your store.

3) Monetary: How much money the customer has spent over a period of time in purchasing the products from your store. A customer who spends more in your store has more likelihood to purchase again from your store as compared to the customer who spends less.

Each customer is scored based on Recency, Frequency, and Monetary values. The higher is the RFM score, the more valuable is that customer as the higher is the likelihood of that customer to respond positively toward the campaign, promotions, etc., and do more purchasing of the products or items from your store again. The order of the RFM score helps in ranking customers accordingly. **Recency** is the most important factor, second is the **Frequency**, and the third is **Monetary**. Recency is considered as the most important factor in RFM model because it tells how recently the customer purchased the products from your store, and it is easier to convince the customer to purchase products from your store more often and spend more by providing lucrative offers. But if the customer has not purchased anything from your store for a long time, it denotes that the customer has less likelihood to return to your store for purchasing and this problem is hard to fix.

RFM Scores Calculation

There are some essential attributes needed for each customer in order to calculate the RFM scores for each customer.[15]

- Most recent date of purchasing by customer.

- Total number of transactions over period by customer.

- Total sales or revenue produced by the customer.

After this the categories number is decided for each RFM attribute. RFM categories numbers can be 3 or 5. In case the RFM attribute is divided into 5 categories, then there will be total of 5*5*5 = **125** different combinations of RFM value starting from the highest **555** and end in lowest **111**. If RFM attribute is divided into 3 categories then there will be a total of 3*3*3 = **27** different combinations of RFM value starting from the highest **333** and ending in lowest **111**. A customer having a higher RFM score of **555** or **333** is generally considered the best customer and more valuable for the business. When the entire data file is scored, there will be a score for each of the three attributes. For the below example, Table 7-1 represents someone who has a score of "5" for Recency, "4" for Frequency, and a "5" for Money and ends up with a score of 4-4-5. Based on the 5 options for each score, you can have 125 unique combinations or segments and similar is the case for 27 unique combinations or segments.

Table 7-1. *RFM*

Recency	Frequency	Monetary	125 Segments
5	5	5	545
4	4	4	
3	3	3	
2	2	2	
1	1	1	

Let's take an example to explain the RFM Model. Looking at the Table 7-2 and considering the details of 5 customers based on their recent, frequent, and monetary purchase. We can score them as customers who have made a recent purchase (in a week's time) are assigned to rank 3, who made purchase in the current month are assigned to rank 2 and who made purchases before 3 months and later are assigned to rank 1. Similarly, for frequency customers who have made purchase more than 10 times are assigned to rank 3, who made purchase between 5–10 times are assigned to rank 2 and who made purchase less than 5 times are assigned to rank 1. Similarly, for monetary customers who spend greater than $1000 are assigned to rank 3, who spend between 500–1000$ are assigned to rank 2 and who spend less than $500 are assigned to rank 1. The last column displays the overall RFM_Score, which is the combination of all the three scores R_Score, F_Score and M_score of each customer. The scoring method depends upon the types of business and Quintile is the most recommended method in most of the scenarios.

Table 7-2. *RFM Score*

Cus_ID	Recency	Frequency	Monetary	R_Score	F_Score	M_score	RFM_Score
1001	2 days	13	1300$	3	3	3	333
1002	5 days	11	1100$	3	3	3	333
1003	26 days	5	600$	2	2	2	222
1004	120 days	1	50$	1	1	1	111
1005	180 days	2	1500$	1	1	3	113

From the above Table 7-2 it is seen that the customers who purchased recently, are purchasing more often and are spending a lot and are assigned the score of 333 Recency – 3, Frequency – 3 and Monetary – 3 are considered best customers. In this example Cus_ID 1001 and 1002 are assigned score of 333 and are the best customers, though Cus_ID 1005 is the highest spender but not considered best customer as that customer purchased only once from your store a long time ago. Cus_ID 1004 is assigned a score of 111 Recency – 1, Frequency – 1 and Monetary – 1, and it means this customer is spending very little, hardly doing any purchasing, and that too was a long time ago. Now the segments can be created based on the RFM scores. In this case we are segmenting our customers into three segments: Best Customers, Potential Customers, and Lost Customers.

- Cus_ID 1001 and 1002 – Best Customers

- Cus_ID 1003 – Potential Customers

- Cus_ID 1004 and 1005 – Lost Customers

RFM analysis helps the direct marketers in understanding customers better and targets them for personalized offers in the future, gains their trust and loyalty, and keeps them happy and satisfied, hence increasing the business revenue.

Overview of K-means Clustering

K-means clustering is a type of unsupervised learning, which is used on unlabeled data, (data when there are no defined categories) and this algorithm was developed by J.B. MacQueen.[16] When in the dataset there are no labels, in such a scenario the task is to find the groups of data in our dataset that are identical to one another and that are known as clusters. This is one of the most popular, simple, and versatile partitional clustering algorithms that can solve the well-known clustering problem and is applicable for any type of grouping. K-means clustering stores k centroids and use them to define the clusters. A point that is nearer to the particular cluster's centroid is assigned to that particular cluster more than any other cluster whose centroid distance is far from that point. In this case study, results of the RFM method is used as inputs for K-means algorithms.

Working of K-means Algorithm

Let $X = \{x_1, x_2, x_3........x_n\}$ be the set of data points and $K = \{k_1, k_2, k_3 .....k_c\}$ be the set of cluster centers. The goal of K-means is to minimize the sum of the squared error over all the K clusters. The main steps of the K-means algorithm are mentioned in the following section:[17]

1. Select the "c" initial clusters centers or centroid randomly.

2. Calculate the distance between each data point and cluster centers (Euclidean distance).

3. A data point whose distance from the cluster center or centroid is minimum is assigned to that particular cluster instead of any other cluster whose cluster center or centroid distance is maximum from that point.

4. Recalculate the new cluster centers or centroids.

5. Recalculate the distance between each data point and new clusters centers or centroids.

6. When the centroid does not change or no data point was reassigned, then stop the iteration or else repeat it from step 3.

Figure 7-4 displays the illustration of a K-means algorithm on a two-dimensional dataset with three clusters. First diagram (Data) displays the two-dimensional input data with three clusters. Second diagram (Seed point) displays the three seed points selected as cluster centers and initial assignment of data points to clusters. Third diagram (First iteration) and Fourth diagram (Second iteration) display the intermediate iterations updating cluster labels and their centers. Fifth diagram (Final clusters) displays final clustering achieved by K-means algorithm at convergence.

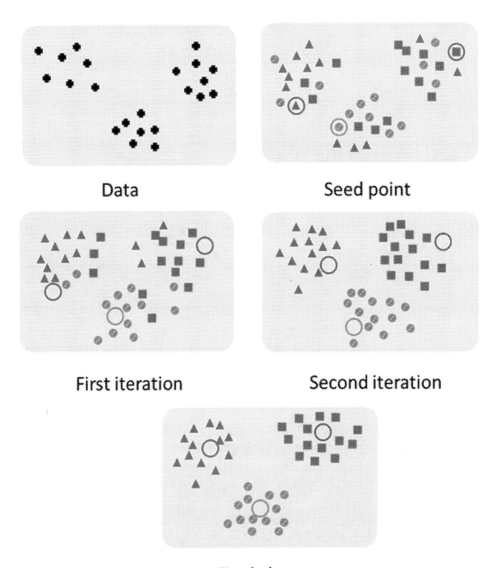

Figure 7-4. *Illustration of K-means Algorithm*

Like all models, a K-means Algorithm has its advantages as well as certain limitations These are discussed in the following section.[18]

Advantages and Limitations of K-means Algorithm:

The key advantages of a K-means algorithm are listed in this section.

- Simple and runs faster on large datasets.

- Easy to implement and interpret the clustering results.

- Fast and Relatively efficient in terms of computational cost.

- Works better and provides the better clusters when data sets are distinct from each other.

A K-means algorithm also has some limitations that are mentioned in this section.

- Identifying the number of clusters K is difficult.

- Sensitive to initial seed selection as initial random seeds produces different final cluster.

- Not very effective in providing better clusters in highly overlapping data.

- Applicable only on numeric data as mean can only be defined for numerical data.

- Sensitive to scale and outliers or extreme values present in the data.

K-means Parameters

In K-means algorithms, there are many parameters that impact the final clustering results of K-means algorithms such as number of clusters K, initialization of cluster, and distance metric. To decide the number of clusters, K is the important criteria in K-means analysis and it is subjective, which mostly depends upon the goals from the clustering process.[19] There are various methods that help in determining the number of clusters in K-means clustering like the elbow method, silhouette method, by rule of thumb, and cross-validation, etc.[20] Another thing is the initialization of a cluster as different initializations produces different final clusterings as a K-means algorithm converges only to local minima. One way to resolve the issue of local minima in

K-means is by running the K-means for a given K, with various different initial partitions and selecting the initial partition with the smallest value of the squared error. An important step in the clustering process is to select the distance metric based on which the similarity of the two elements is calculated. K-means algorithm is typically used with Euclidean distance for calculating the distance between the data points and the cluster centers or centroid. By using the Euclidean distance metric, the K-means algorithms results into spherical or ball-shaped clusters in the data. There is other distance metric other than Euclidean like Manhattan, cosine distance, etc. Different distance metrics produce different clustering results so selection of distance metrics must be carefully done.[21]

RFM Model & K-means Clustering Using R

In this FMCG case study:

> **Business Problem**: Customer Segmentation based on their purchase history
>
> **Business Solution**: To build the RFM model & K-means clustering.

About Data

In this FMCG case study the customer_seg data set is generated synthetically and is used to develop the RFM model and K-means clustering in order to segment the customers based on their purchasing history. The customer_seg data set contains total 330379 observations and 8 variables; 6 variables are numeric and 2 variables are categorical.

```
#Read the data from the working directory, create your own working directly
to read the dataset.

setwd("C:/Users/Deep/Desktop/data")

data1 <- read.csv ("C:/Users/Deep/Desktop/data/
customer_seg.csv",header=TRUE,sep=",")
```

Performing Data Exploration

In Exploratory data analysis, we are going to take a broad look at patterns, trends, summary, missing values, and so on in the existing data. R codes for data exploration and their output are discussed in the section below.

```
#perform exploratory data analysis to know about the data

# display top 6 rows of dataset to see how data look like

head (data1)
```

	Invoice_No	Stock_Code	Product_Category	Invoice_Date	Customer_ID	Amount
1	1540425	154735	Healthcare & Beauty	1/7/2011	556591	23.45
2	1540425	154063	Toiletries	1/7/2011	556591	13.55
3	1540425	153547	Grocery	1/7/2011	556591	14.55
4	1540425	153547	Grocery	1/7/2011	556591	13.55
5	1540425	153547	Grocery	1/7/2011	556591	14.75
6	1540425	154735	Healthcare & Beauty	1/7/2011	556591	23.45

	Country	l_Date
1	United States	12/12/2011
2	United States	12/12/2011
3	United States	12/12/2011
4	United States	12/12/2011
5	United States	12/12/2011
6	United States	12/12/2011

```
# display bottom 6 rows

tail(data1)
```

	Invoice_No	Stock_Code	Product_Category	Invoice_Date	Customer_ID	Amount
330374	1582017	153547	Grocery	12/12/2011	559082	13.54
330375	1582017	154179	Dairy	12/12/2011	559082	26.74
330376	1582018	153547	Grocery	12/12/2011	556391	6.96
330377	1582018	153547	Grocery	12/12/2011	556391	36.00
330378	1582018	153547	Grocery	12/12/2011	556391	237.28
330379	1582018	154179	Dairy	12/12/2011	556391	80.88

```
          Country      l_Date
330374 United States 12/12/2011
330375 United States 12/12/2011
330376 United States 12/12/2011
330377 United States 12/12/2011
330378 United States 12/12/2011
330379 United States 12/12/2011
```

```
# describe the structure of data
```

```
str(data1)
```

```
 'data.frame':      330379 obs. of  8 variables:
 $ Invoice_No      : int 1540425 1540425 1540425 1540425 1540425 1540425
                       1540425 1540425 1540425 1540425 ...
 $ Stock_Code      : int 154735 154063 153547 153547 153547 154735 153547
                       154735 153547 217510 ...
 $ Product_Category: Factor w/ 5 levels "Beverages","Dairy",..: 4 5 3 3 3 4
                       3 4 3 1 ...
 $ Invoice_Date    : Factor w/ 285 levels "1/10/2011","1/12/2011",..: 20 20
                       20 20 20 20 20 20 20 20 ...
 $ Customer_ID     : int 556591 556591 556591 556591 556591 556591 556591
                       556591 556591 556591 ...
 $ Amount          : num 23.4 13.6 14.6 13.6 14.8 ...
 $ Country         : Factor w/ 1 level "United States": 1 1 1 1 1 1 1 1 1
                       1 ...
 $ l_Date          : Factor w/ 1 level "12/12/2011": 1 1 1 1 1 1 1 1 1
                       1 ...
```

```
#display the column name of the data
```

```
names(data1)
```

```
[1] "Invoice_No"     "Stock_Code"       "Product_Category" "Invoice_Date"
[5] "Customer_ID"    "Amount"           "Country"          "l_Date"
```

```
#display the summary or descriptive statistics of the data
```

```
summary(data1$Amount)
       Min.    1st Qu.    Median     Mean    3rd Qu.        Max.
       1.00       6.57     13.65    24.67      22.14   185319.56
```

```
#Let's check the missing values present in the data
```

```
sum(is.na(data1))
[1] 0
```

```
#Unique number of Invoice
```

```
length(unique(data1$Invoice_No))
```

```
[1] 15355
```

The Unique number of invoices in the data is 15355

```
#Unique customer_id
```

```
length(unique(data1$Customer_ID))
```

```
[1] 3813
```

The Unique number of customer_id in the data is 3813

```
#installing dplyr package
```

```
 install.packages("dplyr")
```

```
 library(dplyr)
```

```
#Displaying date format for invoice_date
```

```
data2 <- data1 %>%
mutate(Invoice_Date=as.Date(Invoice_Date, '%m/%d/%Y'))
NOW <- as.Date("2011-12-12", "%Y-%m-%d")
```

```
#Structure of data2 after changing the date format
```

```
str(data2)
```

```
'data.frame': 330379 obs. of  8 variables:
 $ Invoice_No      : int 1540425 1540425 1540425 1540425 1540425 1540425
                      1540425 1540425 1540425 1540425 ...
 $ Stock_Code      : int 154735 154063 153547 153547 153547 154735 153547
                      154735 153547 217510 ...
 $ Product_Category: Factor w/ 5 levels "Beverages","Dairy",..: 4 5 3 3 3 4
                      3 4 3 1 ...
 $ Invoice_Date    : Date, format: "2011-01-07" "2011-01-07" ...
 $ Customer_ID     : int 556591 556591 556591 556591 556591 556591 556591
                      556591 556591 556591 ...
 $ Amount          : num 23.4 13.6 14.6 13.6 14.8 ...
 $ Country         : Factor w/ 1 level "United States": 1 1 1 1 1 1 1 1 1
                      1 ...
 $ l_Date          : Factor w/ 1 level "12/12/2011": 1 1 1 1 1 1 1 1 1
                      1 ...
```

#Building RFM Model

#Calculating Recency,Frequency and Monetary table

```
R_table <- aggregate(Invoice_Date ~ Customer_ID, data2, FUN=max)

R_table$R <-as.numeric(NOW - R_table$Invoice_Date)


F_table <- aggregate(Invoice_Date ~ Customer_ID, data2, FUN=length)

M_table <- aggregate(Amount ~ Customer_ID, data2, FUN=sum)
```

Calculating the Recency, Frequency and Monetary table from the Invoice date and the Amount present in the data:

#Merging the datasets, remove the unnecessary column and rename the columns

```
RFM_data <- merge(R_table,F_table,by.x="Customer_ID", by.y="Customer_ID")

RFM_data <- merge(RFM_data,M_table, by.x="Customer_ID", by.y="Customer_ID")

names(RFM_data) <- c("Customer_ID","Invoice_Date", "Recency", "Frequency",
"Monetary")
```

Recency, Frequency, and Monetary tables are merged together by using customer_id and master table RFM_data is created:

```
#Display top 6 observations from RFM_data

head(RFM_data)
   Customer_ID Invoice_Date Recency Frequency   Monetary
1       555624   2011-01-21     325         1  84902.960
2       556025   2011-12-10       2        88   4014.714
3       556026   2011-12-12       0      3927  40337.760
4       556027   2011-12-09       3       199   4902.968
5       556098   2011-12-09       3        59   1146.574
6       556099   2011-05-12     214         6    115.992
```

```
#RFM scoring

#Rsegment 1 is very recent while Rsegment 5 is least recent

RFM_data$Rsegment <- findInterval(RFM_data$Recency, quantile(RFM_data$
Recency, c(0.0, 0.25, 0.50, 0.75, 1.0)))
```

In this step scoring the RFM data is done by using the quantile method. The scoring is in a range of 1 to 5. In a R segment, 1 is very recent while 5 is the least recent score.

```
#Fsegment 1 is least frequent while Fsegment 5 is most frequent

RFM_data$Fsegment <- findInterval(RFM_data$Frequency, quantile(RFM_data$
Frequency, c(0.0, 0.25, 0.50, 0.75, 1.0)))
```

In Fsegment 1 is the least frequent while 5 is the most frequent score.

```
#Msegment 1 is lowest sales while Msegment 5 is highest sales

RFM_data$Msegment <- findInterval(RFM_data$Monetary, quantile(RFM_data$
Monetary, c(0.0, 0.25, 0.50, 0.75, 1.0)))
```

In Msegment 1 is lowest sales while 5 is the highest sales score.

RFM values and RFM scores are both different things. RFM value is the actual value of Recency, Frequency, and Monetary computed for each customer based on the invoice date and Amount present in the data, while RFM score is a certain number from 1–5 based on that RFM value.

#concatenating the RFM score into a single column

RFM_data$Con <- paste(RFM_data$Rsegment,RFM_data$Fsegment,RFM_data$Msegment)

All the individual scores from Rsegment, Fsegment, and Msegment are concatenated into a single column as Final RFM scores. For example, a score of 144 means Recency-1, Frequency-4, and Monetary-4.

#Total of RFM score

RFM_data$Total_RFM_Score <- c(RFM_data$Rsegment + RFM_data$Fsegment+RFM_data$Msegment)

Summing all the scores from Rsegment, Fsegment, and Msegment.

#Display top 20 observations from RFM_data after scoring

head(RFM_data,20)

	Customer_ID	Invoice_Date	Recency	Frequency	Monetary	Rsegment	Fsegment	Msegment
1	555624	2011-01-21	325	1	84902.960	4	1	4
2	556025	2011-12-10	2	88	4014.714	1	3	4
3	556026	2011-12-12	0	3927	40337.760	1	4	4
4	556027	2011-12-09	3	199	4902.968	1	4	4
5	556098	2011-12-09	3	59	1146.574	1	3	3
6	**556099**	**2011-05-12**	**214**	**6**	**115.992**	**4**	**1**	**1**
7	556100	2011-10-03	70	46	1137.768	3	3	3
8	556101	2011-09-29	74	5	1945.450	3	1	4
9	556102	2011-10-14	59	25	478.832	3	2	2
10	556104	2011-12-10	2	82	1611.692	1	3	3
11	556105	2011-12-07	5	25	526.165	1	2	2
12	556106	2011-12-10	2	56	1227.581	1	3	3
13	**556107**	**2011-01-10**	**336**	**6**	**242.975**	**4**	**1**	**1**
14	556108	2011-11-05	37	38	7575.104	2	2	4
15	**556109**	**2011-03-25**	**262**	**9**	**252.555**	**4**	**1**	**1**
16	556110	2011-11-10	32	27	474.333	2	2	2
17	556111	2011-07-20	145	24	508.118	4	2	2
18	556112	2011-03-05	282	18	379.618	4	2	2
19	556114	2011-10-14	59	175	3224.146	3	4	4
20	**556115**	**2011-06-22**	**173**	**12**	**173.510**	**4**	**1**	**1**

	Con	Total_RFM_Score
1	4 1 4	9
2	1 3 4	8
3	1 4 4	9
4	1 4 4	9
5	1 3 3	7
6	4 1 1	6
7	3 3 3	9
8	3 1 4	8
9	3 2 2	7
10	1 3 3	7
11	1 2 2	5
12	1 3 3	7
13	4 1 1	6
14	2 2 4	8
15	4 1 1	6
16	2 2 2	6
17	4 2 2	8
18	4 2 2	8
19	3 4 4	11
20	4 1 1	6

From the above table it is seen that the **customer_ id 556026 and 556027** highlighted in the italic font are considered the best customers and are assigned a score of 144- Recency – 1, Frequency - 4 and Monetary – 4. Customer_ id 556026 and 556027 are the customers who bought recently, more frequently, and spent a lot and business thrives on such customers. Now on the other side is the **customer_id 556099, 556107, 556109, and 556115** highlighted in the bold font and are considered churned customers and are assigned score of 411- Recency – 4, Frequency – 1 and Monetary – 1. Customer_id 556099, 556107, 556109, and 556115 are the customers who spend the lowest, hardly purchase and that was also quite some time ago. Customers in between with frequent purchases and spending significant amounts of money are considered as potential customers. The RFM technique helps in identifying the high-value customers by understanding their purchase history and behavior. RFM also helps in targeting those customers, especially potential customers, by sending personalized email marketing campaigns in order to increase sales and increase the long-term customer retention and lifetime value.

```
#Display the structure of RFM_data

str(RFM_data)

'data.frame':       3813 obs. of 10 variables:
 $ Customer_ID    : int 555624 556025 556026 556027 556098 556099 556100
                        556101 556102 556104...
 $ Invoice_Date   : Date, format: "2011-01-21" "2011-12-10" ...
 $ Recency        : num 325 2 0 3 3 214 70 74 59 2 ...
 $ Frequency      : int 1 88 3927 199 59 6 46 5 25 82...
 $ Monetary       : num 84903 4015 40338 4903 1147...
 $ Rsegment       : int 4 1 1 1 1 4 3 3 3 1...
 $ Fsegment       : int 1 3 4 4 3 1 3 1 2 3...
 $ Msegment       : int 4 4 4 4 3 1 3 4 2 3...
 $ Con            : chr "4 1 4" "1 3 4" "1 4 4" "1 4 4"...
 $ Total_RFM_Score: int 9 8 9 9 7 6 9 8 7 7...
```

#K-means Clustering

In this case study we used the results of the RFM method as inputs for K-means clustering for determining customer loyalty and better insights about the customer.

```
#to keep only selected variables from RFM data

  clus_df<-RFM_data[,c(3,4,5)]
```

Selecting Recency, Frequency, and Monetary variables from the RFM data for performing the K-means cluster analysis.

```
#Display top 6 observations from clus_df

  head(clus_df)
    Recency  Frequency   Monetary
1     325        1      84902.960
2       2       88       4014.714
3       0     3927      40337.760
4       3      199       4902.968
5       3       59       1146.574
6     214        6        115.992
```

```
#Determining the number of clusters by Elbow method
```

To determine the number of clusters is somehow subjective. Here the Elbow method is used to determine the number of clusters. In addition to the Elbow method there are lot of other methods like Silhouette, Gap statistic, etc., also, which can be used for determining the number of clusters in k-means algorithm. The Elbow method considers the total within-cluster sum of square (total WSS) as a function of the number of clusters. The steps to determine the number of clusters by using the elbow method are explained as follows:

1. K-means algorithm is computed for different values of k. For instance, in this example, the value of k varies from 1 to 15 clusters, maximum value of k is arbitrarily selected but should be large enough to resolve the knee point.

2. The total within-cluster sum of square (total WSS) is computed for each k.

3. Plot the curve of wss versus numbers of cluster k.

4. In the plot the bend (knee) point is generally the indicator of the optimal number of clusters.

```
# Compute and plot wss for k = 1 to k = 15
  k.max <- 15

#Function to compute total within-cluster sum of square
  wss <- sapply(1:k.max,function(k) {kmeans(clus_df,
                 k,nstart = 30)$tot.withinss})

plot(1:k.max,wss,type = "b",pch = 19 ,frame = FALSE,
        main = "Elbow method",
        xlab = "no of cluster k ",
        ylab = "total within cluster sum of square")
```

Figure 7-5 displays the Elbow method. In this plot the bend (knee) point is observed at 3. Hence, in this example the optimal number of clusters is 3.

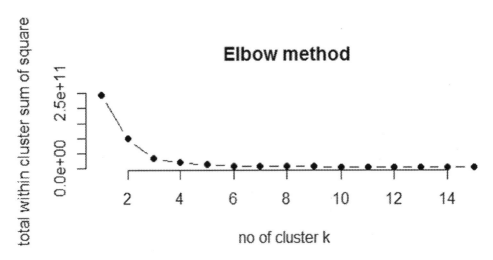

Figure 7-5. *Elbow method*

```
#setting seed

set.seed(123)

#Applying K-means

km <-kmeans(clus_df,centers =5,nstart = 30)
```

The K-means function is used to compute K-means. In this case study we are grouping the data into 5 clusters (centers = 5). As in this case study, RFM model output is the input for K-means clustering. The K-means function has nstart options that are used for attempting multiple initial configurations and selecting the best one. In this case study adding nstart = 30 will produce 30 initial configurations. In this case study, segmenting the customers based on 5 cluster-solutions is recommended as it provides better resolution than after executing the same process for 2, 3 clusters. The decision should be based upon how the business plans to use the cluster results and the level of detail they would like to see in the clusters. Here the business wants to use the cluster results in order to understand the customer behavior and segment them from high–to low value customers.

```
#Describing structure of km
str(km)

List of 9
 $ cluster     : int [1:3813] 3 5 2 5 4 4 4 4 4 4 ...
 $ centers     : num [1:5, 1:3] 2.67 10.62 44.07 92.59 23.92 ...
  ..- attr(*, "dimnames")=List of 2
  .. ..$ : chr [1:5] "1" "2" "3" "4" ...
  .. ..$ : chr [1:3] "Recency" "Frequency" "Monetary"
 $ totss       : num 2.43e+11
 $ withinss    : num [1:5] 2.53e+09 2.84e+09 2.62e+09 2.20e+09 2.68e+09
 $ tot.withinss: num 1.29e+10
 $ betweenss   : num 2.3e+11
 $ size        : int [1:5] 3 37 15 3343 415
 $ iter        : int 5
 $ ifault      : int 0
 - attr(*, "class")= chr "kmeans"
```

K-means output is the list of various bits of information. The most important are:

- **cluster**: It displays the cluster to which each point is allocated.

- **centers**: It contains the values of the five centers computed.

- **totss**: It is known as "Total Sum of Squares" and is defined as the total of the squared distance between a point and the center for the whole data and is computed by adding tot.withinss and betweenss.

- **withinss**: It is known as "Within Sum of Squares" and is defined as the total of the squared distance between a point and its center across all the points in the cluster and it displays the intra-cluster variability.

- **tot.withinss**: It is known as "Total of the Within Sum of Squares" and is computed by adding "Within Sum of Squares" across all clusters.

- **betweenss**: It is known as "Between Sum of Squares" and it displays the intercluster variability.

- **size**: It displays the number of points in each cluster.

  ```
  #Printing km

  print(km)
  ```

Printing km displays the five clusters of sizes 3, 37, 15, 3343, and 415. There are cluster centers (means) for the five groups across the three variables (Recency, Frequency, and Monetary). It also displays the cluster assignments for each data point, for example, 1st data point is assigned to cluster 3, 2nd data point to cluster 5, and so on. "Total Sum of Squares" (totss) and "Between Sum of Squares" (betweenss) measures the goodness of a k-means algorithm. Ideal clusters have the properties of internal cohesion and external separation and the ratio of (between_SS / total_SS should approach to 1. In this case study the ratio of (between_SS / total_SS = 94.7 %) and it indicates a good fit. In other words, high similarity within a group and low similarity between the groups are ideal clusters properties and 94.7% is the measure of the total variance in the data set that is explained by the clustering, hence indicating the good fit.

```
K-means clustering with 5 clusters of sizes 3, 37, 15, 3343, 415

Cluster means:
     Recency  Frequency     Monetary
1   2.666667  250.00000  217852.9513
2 10.621622  487.13514   24271.7027
3 44.066667 1389.06667   66341.6775
4 92.587795   53.49746     937.3011
5 23.918072  269.70361    5953.5805

Clustering vector:
  [1] 3 5 2 5 4 4 4 4 4 4 4 4 4 5 4 4 4 4 4 4 5 4 5 4 4 4 4 4 4 4 4 4 4 4 4 4 4 5 4
 [41] 4 4 4 4 4 4 4 4 4 4 4 4 4 4 4 4 4 4 4 4 2 4 4 4 4 4 5 4 4 4 5 4 4 4 2 4 4 4 4 4 4
 [81] 4 3 4 4 4 4 4 5 4 4 4 4 4 4 5 4 4 4 4 5 4 5 4 4 4 4 4 4 5 4 4 4 5 4 4 4 5 4 4 4 4 4 5
[121] 4 4 4 4 4 4 4 4 5 4 4 5 4 4 4 4 5 5 5 4 4 5 4 4 4 5 4 4 4 4 4 4 4 4 4 4 4 4 4 4 5
[161] 4 4 5 4 4 4 4 4 4 4 4 4 4 4 5 4 4 4 4 5 4 4 2 5 4 4 3 5 4 4 5 4 4 5 2 4 4 5 4 4
[201] 4 4 4 4 5 4 4 4 4 4 4 4 4 4 4 4 4 4 4 4 5 4 5 4 4 4 4 4 4 5 4 4 4 4 4 4 4 4
[241] 4 4 4 4 4 4 4 4 4 4 4 4 4 4 5 4 5 4 4 4 4 4 4 4 4 4 4 5 5 4 4 4 4 5 4 4 4 4 4 4
[281] 4 4 4 4 4 5 4 4 4 4 5 4 4 4 4 4 4 4 4 4 4 4 4 4 4 4 4 4 4 4 4 4 4 5 4 5 5 5 5 4
[321] 4 4 4 4 4 4 4 4 4 4 4 4 4 4 4 4 4 4 4 4 4 4 4 4 4 4 4 4 4 5 4 4 5 4 4 4 4 5 4 4
[361] 4 4 4 4 5 4 4 4 5 4 4 4 4 4 4 4 4 4 4 4 4 4 4 5 4 4 4 4 4 4 5 4 4 4 5 4 5 4
[401] 4 4 4 4 4 4 4 4 4 4 4 4 2 4 4 4 4 4 5 4 4 4 4 5 4 4 4 4 4 5 4 5 4 4 4 4 4
[441] 4 5 4 4 4 4 5 4 4 4 4 4 5 4 4 4 4 4 4 4 4 4 4 4 4 4 5 4 4 4 4 4 4 4 4 4 4
[481] 4 4 4 4 4 4 4 4 4 4 4 4 4 4 4 5 4 4 4 4 4 4 5 5 4 4 4 4 4 4 5 5 4 4 4 4 4 4
[521] 4 4 4 4 4 4 4 4 4 4 4 5 4 4 4 4 4 4 4 4 4 4 4 4 4 4 5 4 4 4 4 4 4 4 4 4 4 4 4
```

```
[561] 4 4 4 4 5 4 4 4 4 4 4 4 4 4 4 4 4 4 4 4 4 4 4 4 4 4 4 4 4 4 4 5 4 4 4 4 4 4
[601] 5 4 4 4 5 4 4 4 4 3 4 4 4 4 4 4 4 4 4 4 5 4 4 4 4 4 4 4 4 4 4 4 4 4 4 4 4 4
[641] 5 4 4 4 4 4 5 4 4 4 4 4 4 4 4 4 4 5 4 4 4 4 4 4 2 4 4 4 4 2 4 4 4 4 4 4 4 4
[681] 2 4 4 4 5 4 4 4 4 4 4 4 4 4 4 4 4 4 4 4 4 4 4 5 4 4 4 4 4 4 4 4 5 4 4 4 4 4
[721] 4 4 4 4 5 5 4 5 4 4 4 4 4 5 2 4 4 4 4 4 4 4 4 4 4 4 4 4 4 4 4 4 4 4 4 4 4 4
[761] 4 4 4 4 4 4 4 4 4 4 4 4 4 4 4 5 4 4 4 4 4 5 4 4 4 4 4 4 4 4 4 5 4 4 4 4 5 4 4
[801] 4 4 4 4 4 5 4 5 4 4 4 4 5 4 4 5 4 4 4 5 4 4 4 4 4 4 4 4 4 4 4 4 4 4 4 4 2 4 4 4
[841] 4 4 4 4 5 4 4 4 4 4 4 5 4 2 4 4 4 4 5 5 4 5 5 4 4 4 4 4 4 4 4 4 4 4 4 4 5 4 3
[881] 4 4 5 4 3 4 4 4 5 4 4 4 4 4 4 5 4 5 4 4 4 4 4 4 4 4 4 4 4 4 4 5 4 5 4 4 4 4 5 5
[921] 4 4 4 4 4 4 4 4 4 5 4 4 4 4 4 4 4 4 4 5 4 4 4 4 5 4 4 4 5 4 4 4 5 5 4 5 4 4 4 4 4 4
[961] 4 4 4 4 4 4 4 4 4 4 4 5 4 4 4 4 5 4 4 4 4 5 4 4 4 4 4 5 4 4 4 4 4 4 4 4 4 4 4 5 5
[ reached getOption("max.print") -- omitted 2813 entries ]
```

Within cluster sum of squares by cluster:
```
[1] 2527929468 2844720962 2623645893 2200791862 2675502481
 (between_SS / total_SS =  94.7 %)
```

Available components:

```
[1] "cluster"     "centers"      "totss"        "withinss"     "tot.withinss"
[6] "betweenss"   "size"         "iter"         "ifault"
```

#Computing centers

km$centers

```
  Recency  Frequency    Monetary
1   2.666667   250.00000 217852.9513
2 10.621622   487.13514  24271.7027
3 44.066667 1389.06667  66341.6775
4 92.587795   53.49746    937.3011
5 23.918072  269.70361   5953.5805
```

Cluster 1 and Cluster 2 can be combined together as these two clusters give the recent purchase group and are considered the **best customers**. Cluster 3 and Cluster 5 can be combined together and considered as **potential customers** and Cluster 4 is considered as **churned customers** as in this group customers spend the lowest, hardly purchase, and that also purchased quite some time ago.

#Assigning clusters to the each data point

km$cluster

```
 [1] 3 5 2 5 4 4 4 4 4 4 4 4 4 5 4 4 4 4 4 4 4 5 4 5 4 4 4 4 4 4 4 4 4 4 4 4 4 5 4
[41] 4 4 4 4 4 4 4 4 4 4 4 4 4 4 4 4 4 4 2 4 4 4 4 5 4 4 4 5 4 4 4 2 4 4 4 4 4 4
[81] 4 3 4 4 4 4 4 5 4 4 4 4 4 4 5 4 4 4 4 5 4 5 4 4 4 4 4 4 5 4 4 4 5 4 4 4 4 4 5
[121] 4 4 4 4 4 4 4 4 5 4 4 5 4 4 4 4 5 5 5 4 4 5 4 4 4 5 4 4 4 4 4 4 4 4 4 4 4 4 5
[161] 4 4 5 4 4 4 4 4 4 4 4 4 4 4 5 4 4 4 4 5 4 4 4 2 5 4 4 3 5 4 4 5 4 4 5 2 4 4 5 4 4
[201] 4 4 4 4 5 4 4 4 4 4 4 4 4 4 4 4 4 4 4 4 4 5 4 5 4 4 4 4 4 4 5 4 4 4 4 4 4 4
[241] 4 4 4 4 4 4 4 4 4 4 4 4 4 5 4 5 4 4 4 4 4 4 4 4 4 4 5 5 4 4 4 4 5 4 4 4 4 4 4
[281] 4 4 4 4 4 5 4 4 4 4 5 4 4 4 4 4 4 4 4 4 4 4 4 4 4 4 4 4 4 4 4 4 4 5 4 5 5 5 5 4
[321] 4 4 4 4 4 4 4 4 4 4 4 4 4 4 4 4 4 4 4 4 4 4 4 4 4 4 4 4 4 5 4 4 5 4 4 4 4 5 4 4
[361] 4 4 4 4 5 4 4 4 5 4 4 4 4 4 4 4 4 4 4 4 4 4 4 4 4 5 4 4 4 4 4 4 5 4 4 4 5 4 5 4
[401] 4 4 4 4 4 4 4 4 4 4 4 4 4 2 4 4 4 4 4 5 4 4 4 4 5 4 4 4 4 4 4 5 4 5 4 4 4 4 4
[441] 4 5 4 4 4 4 5 4 4 4 4 4 4 5 4 4 4 4 4 4 4 4 4 4 4 4 4 4 5 4 4 4 4 4 4 4 4 4 4
[481] 4 4 4 4 4 4 4 4 4 4 4 4 4 4 5 4 4 4 4 4 4 5 5 4 4 4 4 4 4 4 5 5 4 4 4 4 4 4 4
[521] 4 4 4 4 4 4 4 4 4 4 5 4 4 4 4 4 4 4 4 4 4 4 4 4 5 4 4 4 4 4 4 4 4 4 4 4 4 4
[561] 4 4 4 4 5 4 4 4 4 4 4 4 4 4 4 4 4 4 4 4 4 4 4 4 4 4 4 4 4 4 4 5 4 4 4 4 4 4
[601] 5 4 4 4 5 4 4 4 4 3 4 4 4 4 4 4 4 4 4 5 4 4 4 4 4 4 4 4 4 4 4 4 4 4 4 4 4 4
[641] 5 4 4 4 4 4 5 4 4 4 4 4 4 4 4 4 5 4 4 4 4 4 4 2 4 4 4 4 2 4 4 4 4 4 4 4 4 4
[681] 2 4 4 4 5 4 4 4 4 4 4 4 4 4 4 4 4 4 4 4 4 4 4 5 4 4 4 4 4 4 4 4 5 4 4 4 4 4
[721] 4 4 4 4 5 5 4 5 4 4 4 4 5 2 4 4 4 4 4 4 4 4 4 4 4 4 4 4 4 4 4 4 4 4 4 4 4 4
[761] 4 4 4 4 4 4 4 4 4 4 4 4 4 4 4 5 4 4 4 4 5 4 4 4 4 4 4 4 4 4 5 4 4 4 4 5 4 4
[801] 4 4 4 4 4 5 4 5 4 4 4 4 5 4 4 5 4 4 4 5 4 4 4 4 4 4 4 4 4 4 4 4 4 4 4 4 2 4 4 4
[841] 4 4 4 4 5 4 4 4 4 4 4 5 4 2 4 4 4 4 5 5 4 5 5 4 4 4 4 4 4 4 4 4 4 4 4 4 4 5 4 3
[881] 4 4 5 4 3 4 4 4 4 5 4 4 4 4 4 4 5 4 5 4 4 4 4 4 4 4 4 4 4 4 4 5 4 5 4 4 4 4 5 5
[921] 4 4 4 4 4 4 4 4 4 5 4 4 4 4 4 4 4 4 5 4 4 4 4 5 4 4 4 5 4 5 4 4 4 4 4 4 4 4 4
[961] 4 4 4 4 4 4 4 4 4 4 4 5 4 4 4 4 5 4 4 4 4 5 4 4 4 4 4 5 4 4 4 4 4 4 4 4 4 5 5
[ reached getOption("max.print") -- omitted 2813 entries ]
```

#Computing 'Within Sum of Squares'

```
km$withinss
[1] 2527929468 2844720962 2623645893 2200791862 2675502481
```

#Computing 'Total of the Within Sum of Squares'

```
km$tot.withinss
[1] 12872590666
```

```
#Computing 'Between Sum of Squares'
```

```
km$betweenss
[1] 230469229873
```

```
#Computing 'Total Sum of Squares'
```

```
km$totss
[1] 243341820539
```

```
#installing factoextra package
```

```
 install.packages("factoextra")
```

```
#clustering algorithms and visualization
```

```
 library(factoextra)
```

```
fviz_cluster(km, data = clus_df)
```

Figure 7-6 displays good illustrations of the clusters by using fviz_cluster. In this figure, the cluster assignment for each observation is seen. In this example observations 2539, 3235, and 3681 are assigned to cluster 1; similarly observations 3, 1242 are assigned to cluster 2 and so on.

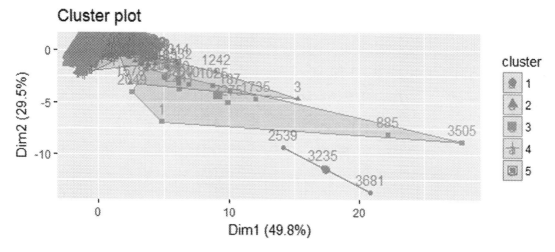

Figure 7-6. *Illustrations of the clusters*

```
#Combining clus_df data and km$clusters by cbind
finalclus<-cbind(clus_df,km$cluster)
```

Cluster output data finalclus is created by combining clus_df and km$cluster where each data point is allocated to each cluster.

```
#Displaying top 20 observations from finalclus
head(finalclus,20)
```

	Recency	Frequency	Monetary	km$cluster
1	325	1	84902.960	3
2	2	88	4014.714	5
3	0	3927	40337.760	2
4	3	199	4902.968	5
5	3	59	1146.574	4
6	214	6	115.992	4
7	70	46	1137.768	4
8	74	5	1945.450	4
9	59	25	478.832	4
10	2	82	1611.692	4
11	5	25	526.165	4
12	2	56	1227.581	4
13	336	6	242.975	4
14	37	38	7575.104	5
15	262	9	252.555	4
16	32	27	474.333	4
17	145	24	508.118	4
18	282	18	379.618	4
19	59	175	3224.146	4
20	173	12	173.510	4

```
#Keeping customer id from RFM_data

cus_id <-RFM_data[(1)]

#Combining cus_id and finalclus by cbind

finalclusid<-cbind(cus_id,finalclus)

#Displaying top 20 observations from finalclusid

head(finalclusid,20)
```

	Customer_ID	Recency	Frequency	Monetary	km$cluster
1	555624	325	1	84902.960	3
2	556025	2	88	4014.714	5
3	556026	0	3927	40337.760	2
4	556027	3	199	4902.968	5
5	556098	3	59	1146.574	4
6	556099	214	6	115.992	4
7	556100	70	46	1137.768	4
8	556101	74	5	1945.450	4
9	556102	59	25	478.832	4
10	556104	2	82	1611.692	4
11	556105	5	25	526.165	4
12	556106	2	56	1227.581	4
13	556107	336	6	242.975	4
14	556108	37	38	7575.104	5
15	556109	262	9	252.555	4
16	556110	32	27	474.333	4
17	556111	145	24	508.118	4
18	556112	282	18	379.618	4
19	556114	59	175	3224.146	4
20	556115	173	12	173.510	4

Customer Segmentation helps in identifying the high valued and potential customers based on their purchasing behaviors. Segmentation enables the marketing team to target the ideal customers with personalized emails and campaigns who are more likely to become leads without wasting money on marketing strategies that will never turn into conversions.

RFM Model & K-means Clustering Using SAS

In this section, we discuss different SAS procedures like proc content, proc means, and proc univariate. We also discuss about building a RFM Model and K-means Clustering for customer segmentation, with an explanation of SAS codes and output of each part.

```
/*Create your own library in SAS like here it is libref and mention the path */

libname libref "/home/aro1260/deep";

/*Importing customer_seg dataset */
PROC IMPORT DATAFILE= "/home/aro1260/data/customer_seg.csv"
     DBMS=CSV Replace
     OUT=libref.synthetic_RFM;
     GETNAMES=YES;
RUN;
/*To check the contents of the data */
PROC CONTENTS DATA=libref.synthetic_RFM;
RUN;
```

Partial output of proc content is displayed in part 1 and part 2 based on their importance in terms of analysis. Proc content displays the information about the data like number of observations, number of variables, library name, and data type of each variable with their length, format, and informat.

Table 7-3.

Part 1

___

The CONTENTS Procedure

Data Set Name	LIBREF.SYNTHETIC_RFM	**Observations**	330379
Member Type	DATA	**Variables**	8
Engine	V9	**Indexes**	0
Created	03/20/2018 14:26:59	**Observation Length**	80
Last Modified	03/20/2018 14:26:59	**Deleted Observations**	0
Protection		**Compressed**	NO
Data Set Type		**Sorted**	NO
Label			
Data Representation	SOLARIS_X86_64, LINUX_X86_64, ALPHA_TRU64, LINUX_IA64		
Encoding	utf-8 Unicode (UTF-8)		

Part 2

___

Alphabetic List of Variables and Attributes

#	Variable	Type	Len	Format	Informat
6	Amount	Num	8	BEST12.	BEST32.
7	Country	Char	13	$13.	$13.
5	Customer_ID	Num	8	BEST12.	BEST32.
4	Invoice_Date	Num	8	MMDDYY10.	MMDDYY10.
1	Invoice_No	Num	8	BEST12.	BEST32.
3	Product_Category	Char	19	$19.	$19.
2	Stock_Code	Num	8	BEST12.	BEST32.
8	I_Date	Num	8	MMDDYY10.	MMDDYY10.

```
/*Descriptive statistics of the data*/
proc means data = libref.synthetic_RFM;
vars Amount;
class Product_Category;
run;
```

Applying proc means, the descriptive statistic or summary of the data is displayed like Number of observations (N), Mean, Standard deviation, Min and Max values of the Amount variable based on the Product_Category are represented.

Table 7-4.

The MEANS Procedure

Analysis Variable : Amount

Product_Category	N Obs	N	Mean	Std Dev	Minimum	Maximum
Beverages	35382	35382	24.0379522	65.8101268	1.2090000	4594.60
Dairy	41412	41412	24.9563357	103.0417944	1.1320000	7196.34
Grocery	204526	204526	24.0829800	466.6886673	1.0011000	185319.56
Healthcare & Beauty	40209	40209	28.3732394	72.5928962	1.1320000	4535.80
Toiletries	8850	8850	22.6455270	84.3323427	1.1320000	3582.84

```
/*Applying proc univariate for detailed summary statistics */

proc univariate data= libref.synthetic_RFM;
var Amount;
run;
```

Proc univariate is one of the procedures that is used to display the detailed summary or descriptive statistic of the data. It will display kurtosis, skewness, standard deviation, uncorrected ss, corrected ss, standard error mean, variance, range, interquartile range, etc. It is also used in detecting the outliers or extreme values present in the data.

Part 1 will display kurtosis, skewness, standard deviation, uncorrected ss, corrected ss, std error mean, etc.

Table 7-5.

Part 1

The UNIVARIATE Procedure			
Variable: Amount			
Moments			
N	330379	**Sum Weights**	330379
Mean	24.6712735	**Sum Observations**	8150870.67
Std Deviation	370.755391	**Variance**	137459.56
Skewness	419.168697	**Kurtosis**	197698.376
Uncorrected SS	4.56147E10	**Corrected SS**	4.54136E10
Coeff Variation	1502.78173	**Std Error Mean**	0.64503199

Part 2 will display more basic statistics like mean, median, mode, std deviation, variance, range, and interquartile range.

Part 2

Basic Statistical Measures			
Location		**Variability**	
Mean	24.67127	**Std Deviation**	370.75539
Median	13.65000	**Variance**	137460
Mode	18.50000	**Range**	185319
		Interquartile Range	15.57500

Part 3 will display the Test column, which lists the various tests like student's t –test, sign test, and signed rank test. The second column is Statistic column, which lists the values of the test statistics; and the third column is p value column, which lists the p values associated with the test statistics. In this case all the tests values and the corresponding p-values are less than 0.0001, which conclude that the variable is statistically significant.

Part 3

Tests for Location: Mu0=0

Test		Statistic		p Value
Student's t	t	38.24814	Pr > ltl	<.0001
Sign	M	165189.5	Pr >= IMI	<.0001
Signed Rank	S	2.729E10	Pr >= ISI	<.0001

Part 4 will display the quantiles like 100%Max, 99%, 95%, 90%, 75%Q3, etc.

Part 4

Quantiles (Definition 5)

Level	Quantile
100% Max	185319.5600
99%	198.9000
95%	67.0000
90%	37.3000
75% Q3	22.1400
50% Median	13.6500
25% Q1	6.5650
10%	3.9240
5%	3.1450
1%	1.9350
0% Min	1.0011

Part 5 will display the five lowest and five highest values of the variable.

Part 5

Extreme Observations

Lowest		Highest	
Value	**Obs**	**Value**	**Obs**
1.0011	203834	7862.19	196163
1.1100	127088	8958.03	81852
1.1320	329406	42868.00	114648
1.1320	324578	84902.96	9468
1.1320	277079	185319.56	330022

```
/*Printing top 10 observations of the data */

PROC PRINT Data =libref.synthetic_RFM(OBS=10);
Title 'Total Sales by Customer';
ID Customer_ID;
Var Invoice_No Invoice_Date l_Date Amount;
Run;
```

It displays the top 10 observations of the data based on Customer_ID, Invoice_No Invoice_Date, l_Date, and Amount.

Table 7-6.

Total Sales by Customer

Customer_ID	Invoice_No	Invoice_Date	I_Date	Amount
556591	1540425	01/07/2011	12/12/2011	23.45
556591	1540425	01/07/2011	12/12/2011	13.55
556591	1540425	01/07/2011	12/12/2011	14.55
556591	1540425	01/07/2011	12/12/2011	13.55
556591	1540425	01/07/2011	12/12/2011	14.75
556591	1540425	01/07/2011	12/12/2011	23.45
556591	1540425	01/07/2011	12/12/2011	24.45
556591	1540425	01/07/2011	12/12/2011	22.14
556591	1540425	01/07/2011	12/12/2011	20.47
556591	1540425	01/07/2011	12/12/2011	19.16

```
/*Creating the Customer_summary table by applying proc sql query */

PROC SQL;
Create table libref.Customer_summary as
select distinct Customer_ID,
max(Invoice_Date) as Recency format = date9.,
(l_Date - max(Invoice_Date)) as Days_since_recent,
count(Invoice_Date) as Frequency,
Sum(Amount) as Monetary format=dollar15.2
from libref.synthetic_RFM
group by Customer_ID;
quit;
```

In the above code, PROC SQL is used for running an SQL query. Create table is used for creating the table as libref.Customer_summary. Select query is used to select the distinct Customer_ID, max (Invoice_Date) as Recency, considering format as format = date9. and creating the new variable as Days_since_recent (difference between l_Date and max (Invoice_Date), count(Invoice_Date) as Frequency and Sum(Amount) as Monetary considering format as format=dollar15.2 from the libref.synthetic_RFM and group by Customer_ID.

```
/*Printing top 10 observations of the data */
```

```
PROC PRINT Data =libref.Customer_summary(OBS=10);
Run;
```

Printing the top 10 observations from the libref.Customer_summary based on the Customer_ID, Recency, Days_since_recent, Frequency, and Monetary value.

Table 7-7.

Obs	Customer_ID	Recency	Days_since_recent	Frequency	Monetary
1	555624	21JAN2011	325	1	$84,902.96
2	556025	10DEC2011	2	88	$4,014.71
3	556026	12DEC2011	0	3927	$40,337.76
4	556027	09DEC2011	3	199	$4,902.97
5	556098	09DEC2011	3	59	$1,146.57
6	556099	12MAY2011	214	6	$115.99
7	556100	03OCT2011	70	46	$1,137.77
8	556101	29SEP2011	74	5	$1,945.45
9	556102	14OCT2011	59	25	$478.83
10	556104	10DEC2011	2	82	$1,611.69

```
/* creating rank */

PROC RANK DATA = libref.Customer_summary OUT = libref.RFM ties=high
group=5;
Var Days_since_recent Frequency Monetary;
Ranks R_score F_score M_score ;
RUN;
```

PROC RANK is used to rank the libref.Customer_summary in group of 5 based on the variables Days_since_recent Frequency and Monetary. After assigning the rank based on the respective variable, new variables are created as R_score, F_score, and M_score; and the output data is generated as OUT = libref.RFM.

```
/* combining RFM */
DATA libref.Final_RFM;
Set libref.RFM;
R_score+1;
F_score+1;
M_score+1;
RFM = Cats(R_score,F_score,M_score);
Total_RFM = SUM(R_score,F_score,M_score);
RUN;
```

```
/*Printing top 10 observations of the data */
```

```
PROC PRINT Data =libref.Final_RFM(OBS=10);
Run;
```

It displays the top 10 observations from the libref.Final_RFM data.

Table 7-8.

Obs	Customer_ID	Recency	Days_since_recent	Frequency	Monetary	R_score	F_score	M_score	RFM	Total_RFM
1	555624	21JAN2011	325	1	$84,902.96	5	1	5	515	11
2	556025	10DEC2011	2	88	$4,014.71	1	4	5	145	10
3	556026	12DEC2011	0	3927	$40,337.76	1	5	5	155	11
4	556027	09DEC2011	3	199	$4,902.97	1	5	5	155	11
5	556098	09DEC2011	3	59	$1,146.57	1	4	4	144	9
6	556099	12MAY2011	214	6	$115.99	5	1	1	511	7
7	556100	03OCT2011	70	46	$1,137.77	4	3	4	434	11
8	556101	29SEP2011	74	5	$1,945.45	4	1	4	414	9
9	556102	14OCT2011	59	25	$478.83	3	2	2	322	7
10	556104	10DEC2011	2	82	$1,611.69	1	4	4	144	9

385

In the below code **PROC FASTCLUS** procedure is used for k-means clustering method. **DATA** names a dataset to use for k-means clustering; in this case it is libref. Final_RFM. The statement **maxclusters=5** inform SAS to form the number of clusters (5) using a k-means algorithm. The **maxiter=100** displays the number of iterations and **converge=0** is used for complete convergence and it means that there must not be any relative change in cluster seeds. The **RANDOM** option displays a simple pseudo-random sample of complete observations as initial cluster seeds, in this case it is **random=121** and **REPLACE** option informs that how seed replacement is made, in this case it is **replace=random.** The statement **OUT** is used to create an output dataset, in this case **out = libref.clust** creates an output dataset as libref.clust and contains the original variables and two new variables, cluster and distance. The variable cluster displays the cluster number to which each observation is assigned and the variable distance displays the distance from the observation to its cluster seed. The **VAR** statement displays the variables used in k-means clustering (Var Days_since_recent Frequency Monetary). In case variable scales differ significantly there is need to standardize the variables, it can be done by using PROC STANDARD. It helps in standardizing variables to a given mean (0) and standard deviation (1). Generally variables with large variances have a larger effect than variables with small variances. In order to have the equal variances standardizing multi-scaled variables is advised, and it must be done before performing the cluster analysis.

```
/*Applying PROC FASTCLUS */

PROC FASTCLUS data= libref.Final_RFM  maxclusters=5  maxiter=100 converge=0
random=121 replace=random out=libref.Clust;
Var Days_since_recent Frequency Monetary;
Title 'FASTCLUS ANALYSIS';
run;
```

The **PROC FASTCLUS output** is split into several sections and each is discussed in the following section.

Part 1 displays the initial seeds table used for each variable and cluster. The first line displays options settings for the Replace= RANDOM, Radius = 0, Maxclusters=5, Maxiter=100 and Converge=0.All these options are set in the PROC FASTCLUS statement. In part 1 output it is seen that a set of points called cluster seeds is selected as a first guess of the means of the clusters. Each observation is assigned to the nearest seed to form temporary clusters. The seeds are then replaced by the means of the temporary clusters, and the process is repeated until no further changes occur in the clusters.

Part 1

FASTCLUS ANALYSIS

The FASTCLUS Procedure

Replace=RANDOM Radius=0 Maxclusters=5 Maxiter=100 Converge=0

Initial Seeds

Cluster	Days_since_recent	Frequency	Monetary
1	23.000000	210.000000	3097.841000
2	14.000000	72.000000	1474.632000
3	29.000000	196.000000	1645.289000
4	32.000000	937.000000	7440.171000
5	245.000000	28.000000	594.445000

Part 2 displays the minimum distance between initial seeds as 211.4824

Part 2

Minimum Distance Between Initial Seeds = 211.4824

Part 3 displays the iteration history and at iteration 39 there is no relative change in cluster seeds; hence convergence criteria is satisfied.

Part 3

Iteration History

Iteration	Criterion	Relative Change in Cluster Seeds				
		1	2	3	4	5
1	4236.6	1.7902	0.9810	1.4893	43.9264	0.9164
2	3876.0	5.4984	0.4170	0.6436	73.9129	0.2171
3	3413.1	7.4046	0.1870	1.1279	114.0	0.1115
4	2905.7	8.7737	0.0689	1.6887	81.1100	0.0520
5	2686.1	12.1396	0.3041	1.8985	73.1816	0.00404
6	2540.4	14.7953	0.5413	2.6382	37.8755	0.0542

(*continued*)

Iteration History

Iteration	Criterion	Relative Change in Cluster Seeds				
		1	2	3	4	5
7	2469.9	16.7801	0.7245	2.9057	13.4303	0.1042
8	2415.8	17.2184	0.8107	2.8022	31.2476	0.1212
9	2351.5	27.6212	0.8411	3.1872	62.8170	0.1353
10	2206.8	38.1209	1.0132	4.0483	102.2	0.1509
11	1925.1	56.7376	1.1160	5.2407	215.6	0.1475
12	1368.1	19.7225	1.3307	5.6947	145.8	0.1550
13	1219.5	16.8525	1.1534	5.1337	0	0.1773
14	1183.6	4.7507	1.2679	4.9083	0	0.1667
15	1165.9	4.6644	1.1813	5.5414	0	0.1372
16	1149.1	0	1.3172	4.8932	0	0.1734
17	1136.2	0	1.1898	3.5941	0	0.1656
18	1129.6	0	1.0010	2.2686	0	0.1417
19	1125.7	4.5520	0.8921	4.1034	0	0.1029
20	1119.1	4.7892	0.7125	3.9832	0	0.0769
21	1111.3	10.8573	0.5340	4.3577	0	0.0613
22	1099.0	12.6014	0.3704	4.1949	0	0.0383
23	1089.3	6.8819	0.5855	4.6327	0	0.0448
24	1082.0	0	0.9912	6.2509	0	0.0652
25	1073.9	0	0.8027	3.7877	0	0.0730
26	1070.6	0	0.7612	3.3173	0	0.0697
27	1068.6	0	0.6095	0.8747	0	0.0751
28	1067.7	0	0.4641	1.8232	0	0.0417
29	1066.3	0	0.5473	4.1132	0	0.0295
30	1064.3	6.6408	0.3372	3.6077	0	0.0334

(continued)

Iteration History

Iteration	Criterion	Relative Change in Cluster Seeds				
		1	2	3	4	5
31	1061.6	0	0.3662	2.3401	0	0.0237
32	1061.1	0	0.2104	0	0	0.0274
33	1061.0	0	0.2700	0	0	0.0345
34	1060.9	0	0.1102	0	0	0.0139
35	1060.8	0	0.0555	0	0	0.00703
36	1060.8	0	0.0837	0	0	0.0106
37	1060.8	0	0.0846	0	0	0.0106
38	1060.8	0	0.0567	0	0	0.00707
39	1060.8	0	0	0	0	0

Convergence criterion is satisfied.

Part 4 displays the criterion based on final seeds as 1060.8. In this case the cluster solution at the 39 iteration is the final cluster solution because the change in cluster seeds at the 39 iteration is less than the convergence criterion. It is also seen from the part 3 output that there is a zero change in the centroid of the cluster seeds for the 39 iteration, which indicates that the reallocation did not result in any reassignment of observations.

Part 4

Criterion Based on Final Seeds = 1060.8

Part 5 displays the summary statistics for each cluster. It displays the total number of observations in each cluster as frequency, root mean squared standard deviation, maximum distance from seed to observation, number of nearest cluster and distance between cluster centroids (distance between the nearest cluster centroid and the current cluster centroid). A centroid is defined as the point that is the mean of all the observations in the cluster.

Part 5

Cluster Summary

Cluster	Frequency	RMS Std Deviation	Maximum Distance from Seed to Observation	Radius Exceeded	Nearest Cluster	Distance Between Cluster Centroids
1	15	7903.7	28169.5		3	42079.7
2	417	1467.6	8740.1		5	5010.8
3	37	5132.3	19649.2		2	18331.5
4	3	20526.1	37953.4		1	151516
5	3341	467.3	2498.0		2	5010.8

Part 6 displays the statistics for each variable. It displays Total SD, Within SD, R-Square, and RSQ/(1-RSQ).

Part 6

Statistics for Variables

Variable	Total STD	Within STD	R-Square	RSQ/(1-RSQ)
Days_since_recent	89.55209	86.63399	0.065091	0.069623
Frequency	208.74803	175.36078	0.295040	0.418521
Monetary	7986	1828	0.947657	18.104771
OVER-ALL	4613	1062	0.947101	17.903837

Part 7 displays the Pseudo F statistic as 17044.45.

Part 7

Pseudo F Statistic = 17044.45

Part 8 displays the Approximate Expected Over-All-R-Squared as 0.95934 (>0.70). Hence this is a good fit model.

Part 8

Approximate Expected Over-All R-Squared = 0.95934

Part 9 displays the Cubic Clustering Criterion as -12.062. The R-square and Cubic Clustering Criterion are invalid for correlated variables. The cubic clustering criterion (CCC) can be used to estimate the number of clusters using Ward's minimum variance method, k-means, or other methods based on minimizing the within— cluster sum of squares. Looking at CCC and increasing it to maximum and increasing the number of clusters by 1, and then observe when the CCC starts to decrease. At that point the number of clusters at the (local) maximum is taken. In case the cubic clustering criterion value is greater than 2 or 3, then it indicates good clusters. Values between 0 and 2 indicate potential clusters and large negative values can indicate outliers.

Part 9

Cubic Clustering Criterion = -12.062

WARNING: The two values above are invalid for correlated variables.

Part 10 displays the Cluster Means of the variables for each cluster. From the above Cluster Means table it is seen that Cluster 3 and Cluster 4 can be combined together as these two clusters gives the recent purchase group and are considered the **best customers**. Cluster 1 and Cluster 2 can be combined together and considered as **potential customers** and Cluster 5 is considered as **churned customers** as in this group customers spend the lowest, hardly purchase, and any purchase was quite some time ago.

Part 10

Cluster Means

Cluster	Days_since_recent	Frequency	Monetary
1	44.0667	1389.0667	66341.6775
2	24.0432	269.0192	5941.5356
3	10.6216	487.1351	24271.7027
4	2.6667	250.0000	217852.9513
5	92.6133	53.4535	935.8017

Part 11 displays the Cluster Standard Deviations of the variables for each cluster.

Part 11

Cluster Standard Deviations

Cluster	Days_since_recent	Frequency	Monetary
1	97.78655	2170.08178	13516.08157
2	33.17629	214.76227	2532.69717
3	29.84064	735.60082	8858.78252
4	4.61880	217.52471	35551.61593
5	91.48950	60.95581	801.91609

CANDISC procedure is used to compute the canonical variables when there are three or more clusters by using the data libref.Clust. The **OUT** statement produces the output data as out=libref.Can. The **VAR** statement is used to display the variables used in the analysis, in this case it is (var Days_since_recent Frequency Monetary). The **CLASS** statement is used for variable cluster as (class cluster) and helps in defining the groups for the analysis.

```
/*CANDISC procedure to compute canonical variables for plotting the clusters */

proc candisc data = libref.Clust out=libref.Can noprint;
var Days_since_recent Frequency Monetary;
class cluster;
Run;
```

In the below code **SGPLOT** procedure plots the two canonical variables produced in **PROC CANDISC**, can1 and can2. The **SCATTER** statements display the variable clusters that define the groups for the analysis and plot the clusters.

```
/*Plots the two canonical variables generated from PROC CANDISC, can1 and can2 */

proc sgplot data= libref.Can;
scatter y=can2 x=can1 / group=cluster;
run;
```

Figure 7-7 displays the plot of canonical variables and cluster value.

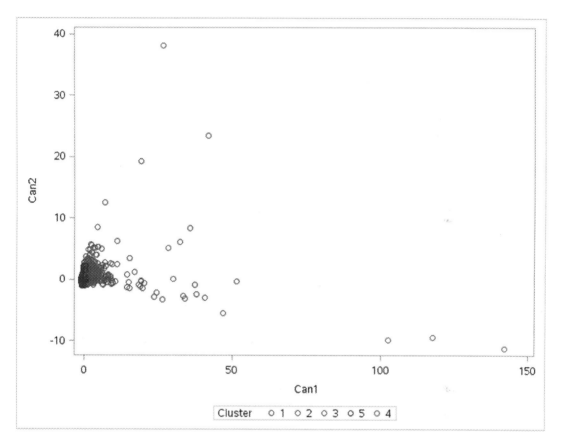

Figure 7-7. *Plot of canonical variables and cluster value*

Summary

In this chapter we learned about application of data analytics in the FMCG industry. The models discussed in the chapter are RFM and K-means clustering. Various characteristics, features, advantages, and limitations of the RFM and K-means clustering were discussed. Practical application of this model in a real-life scenario was demonstrated by help of a case study on customer segmentation based on their purchase history. Model development, execution, and result interpretations were also performed by using both R and SAS Studio.

References

1. Adams, A. The impact of utilitarian and hedonic needs satisfaction on brand trust, brand effect and brand loyalty for selected fast moving consumer goods in South Africa. University of Western Cape, Cape Town, South Africa, 2016.

2. India, T. R. A. Indian FMCG Market 2020; The associated chamber of commerce and industry of India: `https://www.techsciresearch.com/admin/gall_content/2016/11/2016_11$t humbimg102_Nov_2016_004628313.pdf`, 2016; p 42.

3. Research, O. 2017-2022 FMCG Market: Global and United States Regional Opportunities, Competitive Situation, Valuable Statistics and Strategic Analysis; Orbis Research: `http://www.orbisresearch.com/reports/index/2017-2022-fmcg-report-on-global-and-united-states-market-status-and-forecast-by-players-types-and-applications`, 2017; p 101.

4. Javornik, A.; Mandelli, A., Behavioral perspectives of customer engagement: An exploratory study of customer engagement with three Swiss FMCG brands. Journal of Database Marketing & Customer Strategy Management 2012, 19 (4), 300-310.

5. Park, H. The Role of Social Network Websites in Consumer-Brand Relationship. University of Tennessee, Knoxville, `http://trace.tennessee.edu/cgi/viewcontent.cgi?article=2297&context=utk_graddiss`, 2011.

6. Doganis, P.; Alexandridis, A.; Patrinos, P.; Sarimveis, H., Time series sales forecasting for short shelf-life food products based on artificial neural networks and evolutionary computing. Journal of Food Engineering 2006, 75 (2), 196-204.

7. Rakićević, Z.; Vujosevic, M., Focus forecasting in supply chain: The Case study of fast moving consumer goods company in Serbia. 2014; Vol. 10.

8. Perrey, J.; Spillecke, D., Retail Marketing and Branding: A Definitive Guide to Maximizing ROI. Wiley: 2011.

9. Zerman, E. Multi item inventory- routing problem for an FMCG company. Middle east technical university, 2007.

10. Jedermann, R.; Nicometo, M.; Uysal, I.; Lang, W., Reducing food losses by intelligent food logistics. Philosophical transactions. Series A, Mathematical, physical, and engineering sciences 2014, 372 (2017), 20130302.

11. Stănciulescu, G. C., Importance of Reverse Logistics for Retail Acts. In Supply Chain Management - New Perspectives, Renko, S., Ed. InTech: Rijeka, 2011; p Ch. 17.

12. Viglia, G., Pricing, Online Marketing Behavior, and Analytics. Palgrave Macmillan US: 2014.

13. Levinson, J. C.; Horowitz, S., Guerrilla Marketing to Heal the World: Combining Principles and Profit to Create the World We Want. Morgan James Publishing: 2016.

14. Khajvand, M.; Zolfaghar, K.; Ashoori, S.; Alizadeh, S., Estimating customer lifetime value based on RFM analysis of customer purchase behavior: Case study. Procedia Computer Science 2011, 3, 57-63.

15. Wei, J.-T.; Lin, S.-Y.; Wu, H.-H., A review of the application of RFM model. 2010; Vol. 4, p 4199-4206.

16. Li, Y.; Wu, H., A Clustering Method Based on K-Means Algorithm. Physics Procedia 2012, 25, 1104-1109.

17. Jain, A. K., Data clustering: 50 years beyond K-means. Pattern Recognition Letters 2010, 31 (8), 651-666.

18. Marina Santini, Advantages & Disadvantages of k-Means and Hierarchical clustering (Unsupervised Learning). Department of Linguistics and Philology, Uppsala University: http://stp.lingfil.uu.se/~santinim/ml/2016/Lect_10/10c_UnsupervisedMethods.pdf, 2016.

19. Tarpey, T., A Parametric k-Means Algorithm. Computational statistics 2007, 22 (1), 71-89.

20. Kodinariya, T.; Dan Makwana, P. R., Review on Determining of Cluster in K-means Clustering. 2013; Vol. 1, p 90-95.

21. Bora, D.; Anil Kumar Gupta, D., Effect of Different Distance Measures on the Performance of K-Means Algorithm: An Experimental Study in Matlab. 2014; Vol. 5.

Index

A

Affordable Care Act (ACA), 225, 228

Airline industry

air carriers, 277–278

Airport Carbon Accreditation, 279

Air Travel Consumer Report
    November 2017, 284–285

big data, 281

carbon neutrality,
    certification to, 279–280

cybercriminals attack, 283

data, 280–281

defined, 277

flight delays, 284

category, 284

FIM, 286

multiple linear regression (*see*
    Multiple linear regression models)

overall causes of, 285

IATA, 283

offers, 281–282

passenger experience, 278, 281–282

predictive analytics, 281

safety and security, 278

Skytrax, 278

sustainability, 279

working with NASA, 282

Area Under Curve (AUC), 44, 45, 64

Artificial neural networks (ANNS), 234

Assortment planning process, 104

Augmented Dicky
    Fuller Tests, 137–139, 141

Autocorrelation Factor (ACF), 111–112

AutoRegressive (AR) model, 107–109

Autoregressive integrated moving
    average (ARIMA) model

ARMA model, 109

AR model, 107–108

diagnostic checking stage, 113

forecasting stage, 114

identification stage, 111–112

integrated model, 109, 111

MA model, 108–109

Auto replenishment system, 102

B

Bank fraud, 32

Banking sector

cross-selling and up-selling, 29

customer acquisition, 30

customer churn, 30

fraudulent activity, 32–33

loans, 31

Bank-loan defaults prediction,
    logistic regression

assumptions, 38

curve, 37

equation, 35

model fit tests, 39

odds, 36

D. Gupta, *Applied Analytics through Case Studies Using SAS and R*,
https://doi.org/10.1007/978-1-4842-3525-6

Printed in the United States
By Bookmasters